SEA BATTLE GAMES

SEA BATTLE GAMES

By

P. Dunn

Published by
MODEL & ALLIED PUBLICATIONS LTD
13/35 Bridge Street
HEMEL HEMPSTEAD
Herts

In memory of
David Dunn—
the most enthusiastic of
all naval wargamers

Trade Distributors:
ARGUS PRESS LTD.
12/18 Paul Street, London, E.C.2, A 4JS

Printed in Great Britain by Compton Printing Ltd.
London and Aylesbury

CONTENTS

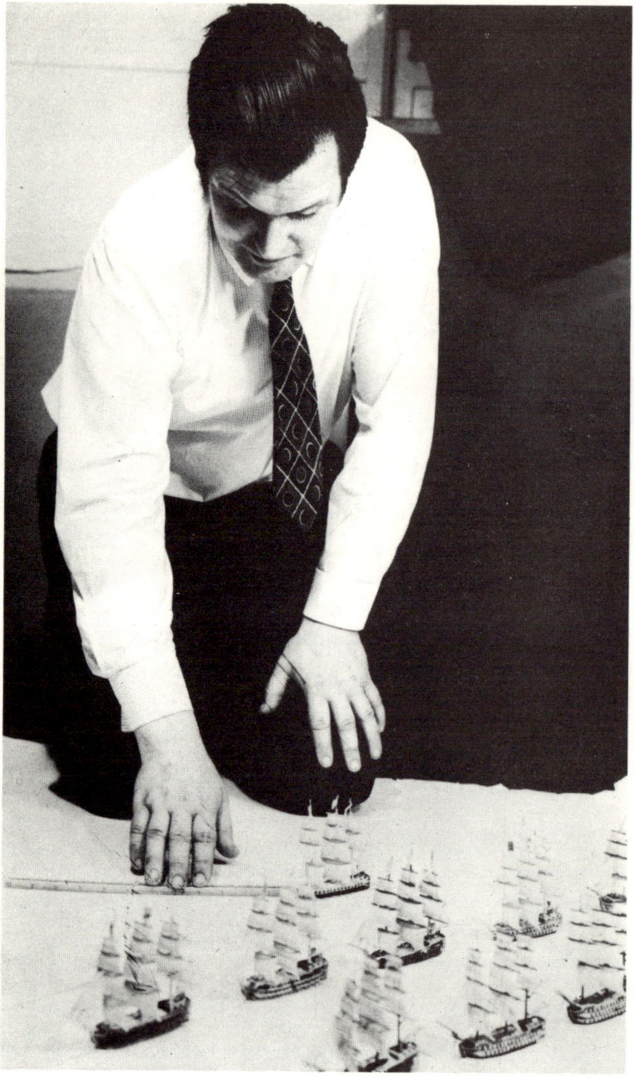

Line of Battle: Settling the issue with gunfire. Masts can be removable to denote damage. The author is placing the ruler at the angle of maximum turn for Spanish ships prior to moving the model.

CHAPTER 1

Background to Naval Wargaming

I do not intend to give a run-down of the history of naval wargaming (or battlegaming as it is sometimes known) because sadly enough, there is very little history attached to this exciting pastime. Our own Fred T. Jane and the American Fletcher Pratt, both famous in naval circles, did bring the hobby into prominence for short periods. Apart from that it appears to have been the pastime of a very few individuals pursuing their activities in relative obscurity.

Credit must be given to that prolific writer of books Donald Featherstone for putting battlegaming on the map. His own magazine and the wargaming books he has written since 1962 have really helped to bring the hobby into prominence. The greatest amount of this interest has, however, centred on the land wargame with model soldiers. Air and sea warfare have remained the poor cousin throughout. Probably less than 10 per cent of all wargamers are fighting ship battles. This figure I confess is merely an inspired guess, but whereas land wargames clubs and magazines are in abundance, there are almost totally lacking similar facilities for the naval battlegamer.

All this is a great pity and almost a tragedy because the key to our past greatness must surely centre on our fighting fleets. I have tried fighting with armies of model soldiers and can attest to the fascination of the hobby. On the other hand I have always preferred to wage naval warfare because I have always found it much more exciting to command fighting ships that deploy armies.

I do know that many naval enthusiasts shy away from naval wargaming because they have formed a wrong impression of the hobby. How can one expect serious-minded adults to enter in such juvenile pastimes! How can one expect to re-create naval warfare with small models and written rules!

I hope this book will show that it *is* possible and that rules can be as simple or complex as your desire for realism dictates. It is certainly not easy and it is a real challenge to get things right. No, it is not kids' stuff to work out rules that really do account for all the important points and do the job properly. This hobby will tax the brains of the most particular of naval enthusiasts. No matter how good the system is, there is always room for improvement and there is really no need for interest to become jaded. Here is one of the fascinations of this pastime. If you have a passion for detail then you are going to be kept busy thinking of a way of building them into a system that works smoothly and without too many tiresome sequences.

Rules need not be complicated to give good results. Even a simple method can give excellent results. We can throw a dice or draw a card to see the result of an entire broadside or you can throw a hundred dice to see the damage caused by a single shot. The choice is yours.

It is to be hoped that the rules presented herein will give the newcomer much pleasure and excitement and that he will find his venture into the field of naval battlegaming a rewarding enterprise.

It is not just fighting battles that is so pleasurable. The attendant model making, research, etc., is also enjoyable. The biggest success that this book can enjoy is if it attracts the reader to the hobby. Even if he does not use a single system presented herein but is inspired to try and better them, then one of the purposes of the book will be fulfilled.

I do sincerely believe that naval wargaming could be enjoyed by all and not just those interested in warships and naval history. I have press-ganged several unwillings into taking part and they have thoroughly enjoyed themselves.

Fletcher Pratt's great games were the same. Hundreds of people took part. Most of them came merely out of curiosity to see what was going on. When they saw the impressive fleets of model ships and sensed the electric atmosphere of excitement and enthusiasm, they wanted to take part. Most of them knew little or nothing about naval warfare. Soon, though, they were anxious to learn more and the very nature of the activities made them experts on ship fighting and interest in warships in general followed.

The author admits that he tried a similar experiment and failed. The bulk of those who joined from the onset were youngsters. Adults and juniors do not really mix together in such activities for fairly obvious reasons. If adults are to take this past-time seriously they want to congregate with their fellows. The club then became almost exclusively for young members and they did not want to throw open the games to the public. It therefore became confined to naval enthusiasts and membership never exceeded more than twenty.

To emulate Pratt's great games in this country, the whole thing would need to be organised by and consist mainly of adults. This book, of course, is written for the individual wargamer and assumes that the reader will pursue his hobby in company with a friend or very small group. The next step though is to form a club proper and a very rewarding step this might prove to be. You will need to decide whether your club or society will be an exclusive affair or one in which the public at large are welcome to come along and take part. The biggest problem is in formulating a method of play so that everyone can participate together and long periods of inactivity by sections of those taking part are eliminated.

This then is the unlimited horizon that stretches before those who care to take up the hobby. You can keep it a simple affair between two players or expand into a club with unlimited membership. Not only is there a great variety of periods to choose from or change to, we can never exhaust the possibilities of tackling each one in a new or improved fashion. I hope that you will make a start today.

CHAPTER 2

Scope of the Subject

When this book was first planned, it seemed logical to write chapters about all the most important aspects. It seemed necessary to write about weather, morale, the problems and solutions of scale and so on.

Then I decided that these things are self-evident and that the reader will quickly learn how to tackle the problems himself. It is earnestly hoped that these pages will indicate how it can be done. He will see the principle involved in rule making and write his own rules for those things that *he* thinks are important and which I have not covered. He may even decide he can do a better job all round!

What about the absolute beginner who may know little or nothing about ship fighting or related subjects? I stick by my maxim. When he takes up this pastime, he will obviously stage something simple at first. This will be followed by a desire to improve his game. Interest will deepen and research will follow. The beginner will soon become experienced and he will have definite ideas of his own. He does not want to be told what he must and must not do by any so called experts. The truth of the matter is, we are all learners. The subject is vast and the horizon is very wide indeed. If anyone should tell me that he has invented the ultimate set of rules, I should be inclined to smile. Something better or more realistic is always waiting to be invented.

If an attempt was made to discuss everything related to naval battlegaming, a truly vast volume would result. Try to imagine for a moment the thousands of different points that would have to be covered if we attempted to reproduce the Pacific WW2 type warfare in detail. Since the wargamer is limited to the amount of complexity he can handle, much of the contents of our major work would be unnecessary data.

No, I am firmly convinced that this book should be nothing more than an insight into the art of naval battle-gaming. For those who want ready-made rules, they are here. Those whose main desire is to learn something about the subject so that they can formulate their own rules, will also be satisfied I trust.

Now what of the actual content of this book? There are three complete fighting systems for the Napoleonic period, 1870 ironclads, and WW2. I have also introduced a fairly complex gunnery system for pre-Dreadnought type warfare. It seemed obvious that enough was said about manoeuvring in the rules for 1870 and WW2 to enable the reader to modify either to suit the period between the two. The Second World War rules are easily adaptable to suit WW1 once allowances are made for inferior torpedoes, gunnery ranges, different techniques for submarine attack, etc.

I have also shown how battles can be fought without models and how secrecy and whole campaigns can be included. There are also complete rules for fighting a miniature 'world war' campaign.

The Second World War rules are not supposed to be highly technical or super realistic because the aim is to allow each player to handle large forces. All the most important features of ship fighting are covered and thus the use of detailed battle systems and complex rules is not practical.

Why have only four periods been covered in this book? The answer to that is that if an attempt were made to cover more, then the rules would be too sketchy to be of much use. A few readers may be disappointed to see that eras of naval warfare prior to the mid-eighteenth century are not included. There is a definite reason for this. It seems pretty certain that the bulk of naval enthusiasts prefer the age of steam with all its mechanics, science, and specialised side effects. I have therefore merely written a short chapter loosely to discuss these early periods and hint at how the battlegamer should tackle them. The reader may be interested to see that the furthest point in time I have covered is the year 1470. This is the first time that guns were ever mounted aboard ship and we note that this happened exactly 500 years ago!

Now a friendly word of warning for the aspiring rule maker. Probably the greatest menace to historical accuracy were the well meaning Victorian writers. The splendid deeds of Nelson had surrounded the Royal Navy with an aura of invincibility. The British empire was then at its greatest and the Victorian historians, with patriotic pride in their breasts, wove romantic stories around the deeds of earlier Englishmen—Alfred, Drake, and many others, became absolute supermen. The historic truth of the Spanish Armada is, alas, far removed from the popular Victorian legend.

So research is important. This need not be of an intense or technical nature but do sort out fact from fiction. Be very careful of books published by Victorian authors. There *are* exceptions of course. The works of Sir William Laird Clowes, for example, were excellent books.

Another point about research. Your fact finding should take you to the battle stage. Ignore what weapons were supposed to do or what the manufacturers intended. Ignore the sometimes impressive but often misleading results of test range firing. The only thing of any use to the battlegamer, is what the weapon could do in practice. To take one modern example first. The Americans developed a highly sophisticated rocket to home on to Vietcong radar stations and destroy them. It seemed a foolproof weapon and any 'modern' wargamer would have placed a high value upon it. The Vietcong answer was to simply switch off their radar sets when the rockets were airborne, thus making this impressive weapon entirely useless!

Another case that can be mentioned is the English culverin that we used against the Armada in 1588. It was supposed to destroy the Spanish galleons at long range but they entirely failed to do themselves justice. It took actual battle conditions to learn the bitter truth about the gun.

Now I should like to mention the biggest headache to be encountered in rule making. This headache is the unforeseen and the unpredictable. It is all too easy to sit down and write out what appear to be the perfect set of rules. For some mysterious reason the rules seem to utterly fail to give the desired results and countless situations occur that were completely unforeseen.

I can remember spending a week developing a new gunnery system. I reviewed it thoroughly and tested it out on paper and thought it perfect. When, however, it was first used, this system gave absolutely terrible results.

Wargames are always plagued with situations cropping up that we could not possibly foresee and for which no rules exist to cover the situation.

The answer is that a really satisfactory set of rules takes ages to develop. It is all worth while when they are eventually 'perfected' but there can be a lot of frustration meantime. So take your time. Do not attempt anything over ambitious at first. It is better to start with a simple set of rules and then add to them and improve them as faults and omissions reveal themselves.

The naval wargamer really has an imposing task. He has to re-create in written rules all the positive science of naval warfare. Each class and type of warship has a potential offensive and defensive capability that is somehow different from its fellows. The wargamer has to account for this. He has to duplicate its importance in battle and give it the power that it really had. If he makes *Anson* more powerful than the *Iowa* then his rules are faulty. Not that *Iowa* could not be defeated by the lesser British battleship, rather that all things being equal, the American ship should win. The fortunes of war though, are well known to all.

Then again, we must try to get things relative. It is no good merely making a Battleship twice as powerful as a Cruiser or ten times as powerful as a Destroyer. We must assess the firepower and defensive quality of each and get a degree of realism in our rules. It is true that in real life, if the same battle could be exactly repeated, it is likely that entirely different results would happen the second time. On the other hand, there are certain things that *cannot* happen in either battle (i.e. Destroyers sinking a Battleship with gunfire) and your rules should follow suit.

The true secret of rule making is to account for as many factors as possible with the less complicated method you can devise. The WW2 gunnery table is an example. It accounts for number of guns firing, range to target and calibre of gun firing. Only one card is required to compute the result. It would have been easier to invent a system when 3 or 4 cards are cut for the result but it would become tiresome to keep drawing them.

In the course of reading this book, you will see that I am an advocate of the campaign. The set-piece is good fun and a logical beginning. Its drawback is that the battle occurs for its own sake. There is no real issue at stake. The result of this is that unrealistic situations arise. We find the losing side determined to go out in a blaze of glory. He does not attempt to escape with the remnants of his fleet, preferring to throw everything into the scrap and fight on until not a single ship remains afloat. How untrue of real war!

I have seen battlegamers with casual indifference, write off damaged battleships by 'charging' the enemy fleet. Yet because it takes tremendous skill, effort, time, and money to build one of these monsters, real world powers were only too anxious not to take risks with them.

The campaign is something quite different. The loss of a heavy unit is a step towards defeat of the whole war effort. We are far less inclined to risk our big ships and are more anxious to preserve them than see them go down fighting gloriously.

Special steps can be taken to make a set-piece a strategical affair and something more refined than a mere 'blast' battle, but it can never be as truly realistic as the map campaign. Naval warfare is not just warships shooting it out, it is a hundred other things as well. Battles usually happen because of important issues. X wants to invade Horra island. Z wants to stop him. There are a few exceptions. The French had no need at all to fight the battle of Trafalgar for example.

The biggest satisfaction we derive from this hobby is when we prove our tactical skill and defeat our opponent. As far as the campaign goes, this skill is not confined to battle. The way we plan and conduct our operations is a vital factor too. We, who in real life are hardly likely to see above the lower deck, are now in command. It is us that are the great brains behind the war effort. It is our tactical genius that can rout a superior enemy force. Battlegaming is a wonderful short cut to the admiral's bridge. We do not need to spend thirty or forty years getting there. Yet it is quite true to say that provided our rules faithfully reproduce the manner in which the ships of our chosen period manoeuvre and fight, the same opportunities abound to emulate or even outdo Nelson, Tojo, and Cunningham.

The same tactics or situations open to a real admiral are likewise open to yourself. You can try anything that you care to dream up. The standard line of battle is the safest tactic but there is nothing to stop you trying something new. Nelson's plan of attack at Trafalgar was on the face of it both lunacy and highly dangerous. Great admirals have always taken risks. Nelson's genius was in the fact that he would attack the enemy under circumstances that no one else would dare do with twice his force.

In keeping with the subject, let us briefly analyse the two greatest naval victories in history. These two are Trafalgar in 1805 and 1905 at the battle of Tsu-Shima.

Why were such crushing victories scored by England and Japan? The answer is that in the first battle the French and Spanish fleets were badly inferior in gunnery and seamanship. To add to this, they had to contend with Nelson who knew how to take advantage of his enemies' failings.

In the second instance exactly 100 years later, we find two fleets evenly matched in skill but one of them commanded by an admiral who did not know his business at all. He tamely allowed his opponent to outmanoeuvre and destroy his own fleet.

The Russian admiral Rozhestvensky has a certain claim to

fame because of the remarkable achievement of steaming his fleet 18,000 miles to do battle. During the actual battle, he did not make a single tactical decision or attempt to counter or even follow the manoeuvres of the famed Admiral Tojo. The Russian gunfire had been highly accurate when the battle commenced but the tactical situation became so intolerable that morale plunged to near zero and they hit nothing at all thereafter. The entire huge Russian fleet was sunk or captured.

Thus we may observe that the magnitude of the French/Spanish fleet's defeat was due to the failing of fleet training. It can hardly be blamed on Admiral Villeneuve, whose own tactics were perfectly sound. Nelson's method of attack could not have succeeded against an enemy as well versed in gunnery as his own fleet.

The failings of the Russian fleet can be laid upon the head of its commander. An efficient admiral such as Makarov would at least have made the task of Tojo immensely more difficult.

If two fleets are both well trained and commanded by a competent admiral, then spectacular results are unlikely. Why was the great battle of Jutland in 1916 so indecisive? Because the German fleet was as efficient as the British and because the admiral really knew his job. Therefore we should be able to see that the result of any battle will depend on the quality of both fleets and the quality of the two admirals. The battlegamer should also bear in mind that Nelson shattered his foes because of his revolutionary new tactics and willingness to battle any odds. Such things tried by an aspiring admiral lacking the master's genius, may have disasterous results. When the French copied Nelson's tactics at the first battle of Lissa, their fleet was shattered.

Thus we can take advance warning. The most exciting and closely fought actions will be between two players whose skill is evenly matched. It is doubtful if one side will gain a great victory unless one player is badly outnumbered. In any event the point is this. To win and lose is not merely a matter of luck—although it is bound to influence the outcome. The final result will depend on you and the way you handle your model ships.

To conclude this introduction, I have prepared a short list of some of the points of research that the reader will need undertake when he is attracted to writing rules for a specific period.

1. The different type of vessels then in use and the purpose for which their design was intended or used.
2. The sizes of these vessels and the number of crew carried.
3. The general quality of the fighting ability of the countries involved.
4. The sailing quality and manoeuvrability of the vessels.
5. The average and maximum speeds of various types (or specific speeds if known).
6. Fighting method. Is boarding and ramming possible? If so, how difficult was it to do either.
7. How much bad weather will affect the ships.
8. How long can the vessels remain at sea. Many ancient vessels were strictly coastal.
9. If missiles are fired, what is likely to be the damage caused. Missiles include arrows, spears, rocks (from war engines) and gunfire.
10. Weapon capability. Ranges. Odds of hitting. Rate of fire. Rounds carried aboard. Damage potential.

This is the sort of thing we need to know for the typical wargame. The stickler for realism may want to know a lot more.

All this relates to the ships. Campaign rules demand even greater research.

CHAPTER 3

Models for Wargaming

Those who fight battles with model armies do not have many problems when it is desired to create a tabletop army. A quick visit to the local model shop and we can return home armed with six packets of excellent little Airfix soldiers. We have spent less than one pound and now have two sizeable forces consisting of infantry and cavalry, complete with cannons. Our only chore is to paint the figures. The land battle-gamer never has to 'build' his own figures.

The naval wargamer can create a Napoleonic fleet in the same way by simply buying the small Airfix *Victory* kits and converting them to represent other vessels as well. Rather *more* effort is involved because the kits have got to be assembled and then cut down and finally painted. The cost, however, is slight—just a few shillings for each vessel.

Outside of this particular period though, things are more difficult for us. Small models for earlier periods hardly exist and though a truly staggering range of models are made for the age of steam, problems of cost are involved. A single vessel of cruiser size will cost from 25s. to £3. More of these models later.

The cheapest way to provide a model fleet is fairly obvious. We shall have to make the ships ourselves. This will appear a daunting task to those who have never before attempted to scratch-build a model warship. It need not be, because a simple block model should not be beyond the capability of any of us. The author confesses that model making is not one of his hobbies. Yet you will see that his first attempt to scratch-build a simple block model resulted in something that is quite good enough for wargaming purposes. The hull is simply two layers of balsa wood and the superstructure is

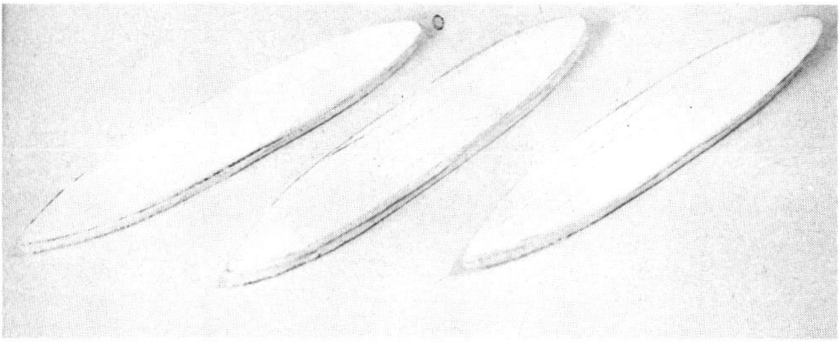

Simple ship shapes. The easiest way to acquire a 'model' fleet. Superstructure and turrets or masts and deck guns can be drawn on. Different sizes and colours denote vessels of varying importance or type.

merely layers of those rubbery type domestic tiles cut into rough shape and size. Pins represent ventilators and the funnels are cut from any piece of wood that happens to have the right proportions. Thin pencils and cocktail sticks are one source. Most of the smaller detail is just painted on and thick coats of paint hide a multitude of other sins. The models, though, do look like the ships they are supposed to be (American semi-dreadnoughts) and each one was made in an hour or two.

Provided the funnels and superstructure are roughly cor-

Simple block models made by the author. They look more effective when viewed on the floor than under the critical eye of the close-up camera. This sort of model can be made in an hour or two.

Models not yet available from commercial sources and made for the author by individual model makers. Scale 1–1200, wood and plastic.
The big ship is a Japanese '13' class battleship with 18-inch guns. It was never completed. Built by Ron Hughes of Nazeing.
The other vessel is an early Russian dreadnought, built by John Trotter of Whitby, Yorkshire.

rect, it is surprising how realistic a simple model can look.

The most rapid and effortless way to prepare a model fleet is to simply cut the shape of the vessel from balsa wood or card. This is how a seagull would view the vessel as it hovered directly overhead. The big ships are readily identified by their size and smaller vessels of roughly equal size (i.e. frigates and corvettes) can be recognised by various colour schemes and numbers. Apart from the lack of visual effect, such 'models' will prove entirely satisfactory and each can be made in a matter of minutes or even less.

Another easy scheme is to draw the silhouette of a warship and mount the drawing on a card. The card is slotted into a simple stand and the whole thing serves the purpose of a model. The silhouette can be drawn on both sides of the card if desired. If these 'flats' are to represent modern warships (1880-1945) then make the drawings as small as possible. It is possible to have your drawings or silhouettes from a book photo-copied. This is useful when there are many vessels of the same class. There were 174 Fletcher class destroyers for example.

The silhouettes shown in this book are reductions from original drawings that were much larger. When fine detail on a small drawing is required, a large drawing has to be made and then reduced in size by a special photographic process. Simple outline drawings will be all that are required in most cases.

A good idea that was put to me, was to use little coloured tabs to slot on top of the ship cards. Blue might denote the amount of fuel remaining and a red tab might indicate damage suffered, etc. The tabs are obtainable from office shops and some stationers. Their true purpose is to be attached to files and have various indications relevant to the content of the files. As battlegamers, we can put them to other uses.

Models from commercial sources

Throughout this book there are photographs of commercial warships. All the metal models are made to the international recognition scale of 1-1200 or 1-1250. Because of the enormous range available they are widely used as wargaming pieces. If, however, you decide to build your own, you would be well advised to make them to a smaller scale. A model of the *Littorio* (Italian battleship) would be 7 ins. long and if scale was kept constant, her guns would fire 1,200 ins. and she would move over 30 ins. every minute. Of course we can make reductions in scale but there is a limit as to how much we can reduce these things.

Let us look at the firms that make these 1-1200 models (or 1-1250).

Triang (British). Models difficult to obtain as the range went defunct years ago. Battleship *Vanguard* cost only 7/11d. and some British cruisers and destroyers were also made.

Eaglewall (plastic). You will be lucky if you find any of these as collectors snap them up. They were very cheap. WW2 vessels. British and German.

Pyro (plastic). Still available in some shops. American and Japanese battleships and carriers. American firm.

Superior. These are metal and very robust. An extremely large number are available and are mostly warships and auxiliaries of WW2. Some early ships are also manufactured. Average cost of a battleship is about 35/-. English agent is J. Holton, 2 Aragon Avenue, Ewell, Epsom, Surrey. Price list costs 1/6d.

By far the largest makers of these small models are the Germans. Over 1,100 models (not all warships) are made between them. Here is a list of these firms and some notes on the models they manufacture.

Star. American Civil War and a mixed bag of other ships

More models made for the author's collection. The cruiser is a late war American 8-inch-gunned ship. Made by Mick Earll. Note her size compared to the French semi-dreadnought *Danton* of 1909. Model built by Barrie Richardson of Sutton-at-Hone.

The German firm Navis make 270 different models and bring out new ones each month. Here are twenty-two of them. The cruiser *Koln* is a Hansa model and the light grey destroyer was made by Ron Hughes. Navis specialise in ships afloat during World War 1.

from 1921-61. The models can be purchased in kit form or ready made.

Trident. All nations, 1921 to present day. Many Japanese WW2 ships.

Anka. Details unknown. May be an American firm. Models are extremely expensive and are probably too detailed for wargaming.

Delphin. Mixed bag of warships from 1922 to present day. Many Japanese and American WW2 ships.

Hansa. Another mixed bag. Many German and American ships of WW2. Some early ships and many post-war models too.

Neptun. At present this firm lists the whole of the Japanese WW2 fleet of larger warships. Some American WW2 vessels are also made.

Copy. Limited range includes II British WW2 or earlier vessels. Also French and Italian ships and two American models. 1921 to 1942.

Schlingelhof. A few plastic models including an American 'jeep' carrier.

Atlantic. Nineteen of the twenty models listed are post-war and none are warships.

Wiking. A mixed bag of vessels mostly post-war. Their pre-war range of models are also available at about half the cost of models made from recent moulds by other firms. These are all warships.

Navis. This firm specialise in ships afloat during WW1. About 270 different models are made and represent the fleets of most nations.

Mercury. Details unknown but thought to be mainly freighters, etc.

Mercator. Early warships are the speciality of this firm. There is a good selection and the Russian fleet in particular is well catered for. The average period covered may be said to be about 1890.

Although a ready-made source of wargaming models, it should be pointed out that many of them are expensive.

Superior Models: Superior are an American firm and make an extremely large range of warships and auxiliaries. Most are of World War 2 period. Shown here are just a few of the American fleet.

Some battleships cost up to £5 each. Many of the models are super detailed and might be considered too good for wargaming as a knock can damage them.

As an indication of cost, I have listed one model battleship from each firm. They are all dreadnoughts. Price is given in German marks. At the time of writing, a mark is valued at 2/4d. When sold in England, heavy purchase tax is added. These prices are what the Germans pay.

Firm	*Model*	*Cost*	*Model tonnage*
Copy	*K.G.V.*	20.00	35,000
Delphin	*Nelson*	26.40	33,000
Hansa	*Hood*	30.80	42,000
Mercator	*Jamie I*	16.00	16,000
Navis	*Revenge*	20.00	27,500
Neptun	*Kongo*	29.00	32,000
Star	*Minas Gerais*	11.00 (kit)	20,000
Trident	*Akagi* (carrier)	33.50	36,500
Wiking	*Nelson*	9.80	33,000

All the battleships listed were built in the British Isles except for *Jamie I*.

The models can be obtained direct from German wholesalers but it is rather a complicated procedure. They send you an invoice for the models you want and you pass this on to your bank who will pay the bill for you. When your wholesaler receives the money he posts the models and the postman collects the tax outstanding which is due on the models. You cannot pay by cheque or postal order and cash cannot be posted abroad. It is possible to pay by international money order.

Hansa models are sold in Hamley's toyshop, Regent Street, London. Delphin models are sold by Greenwood & Ball, The Garrison, 198 Northolt Road, South Harrow, Middlesex. Price list and catalogue 2/-.

Hamleys toyshop also stock the Avalon-Hill wargames

Scale: 1 inch = 2,000 yards. Old positions erased.
Method (1) Drawn upon paper.

(American). Some of the naval games are *Bismarck*, *U-Boat*, *Midway* and *Jutland*.

Models can also be built to order. You will see that there are photographs of 1-1200 scale models built by individual model-makers in this book. The price they charge you will vary according to your needs. A really first-class model of a battleship will cost anything from £3 to £5. Considering the work and time involved, this is not excessive and if the specifications are less exact they will not cost so much.

These model makers are not professional people as a rule. They make models for themselves and their friends, but will usually be pleased (time permitting) to make some for other people. Since such people without fail take 'Model Boats' every month, your needs expressed in the classified advertisements will probably bring you a reply.

Catalogues. Those readers requiring catalogues should send two international reply coupons (get them from a large post office) to either of the German wholesalers listed below. These list the entire range of models.

Method (2) Simple ship shapes made from wood or card.

Egon Wiedling. 8 Munchen 2, Theatinerstr 13. (airmail letter 9d.).

Christian Schmidt. D 8000 Munchen 55. Sauerbruchstrasse 10.

Tony James of London is a well-known wargamer and collector who has made up his own list of models for the benefit of other collectors. A s.a.e. with initial enquiries should be sent to A. C. James, 97a Kirkwood Road, London, S.E.15.

Other commercial models

I have already mentioned the Airfix *Victory* kit. This is sold in packets (not boxes) and is roughly 1-600 scale. There are other kits in this range but they are not to a constant scale. The kit of H.M.S. *Shannon* (1812) is the same size as

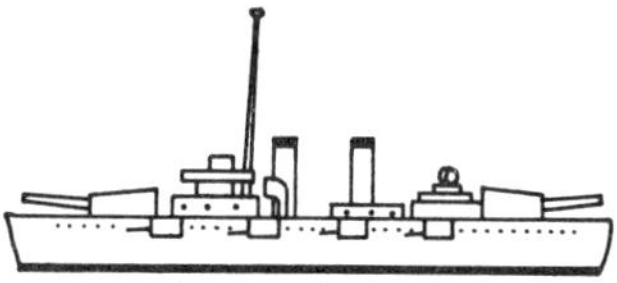

Method (3) Simple block model.

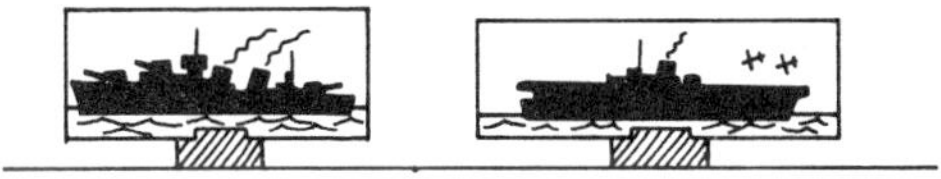

Method (4) Ship Cards.

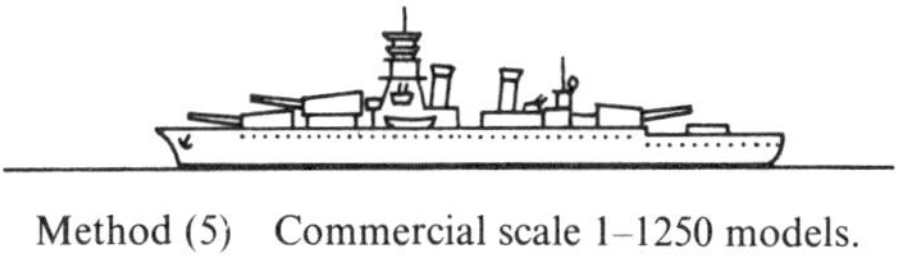

Method (5) Commercial scale 1–1250 models.

Victory, though she is only a frigate. Also made in this range are *Revenge, Golden Hind, Mayflower* and several others.

The *Victory* kit is of special interest because it can be converted quite easily into larger and smaller vessels. A really large wargames fleet can be built up at less than 3/- per model and the size of these little kits are just right (see photographs).

As soon as the hull sections are glued together, it is necessary to cut off the base of the hull with a hot knife. A very jagged edge will result unless great care is taken and it is all too easy to burn not only the model but your own hand as well. The method is very much quicker than sawing though. Where you cut the waterline will depend on whether you want a two-decker or three-decker.

To lengthen the model you need two hulls. Cut off a larger section of the bow on one hull than on the other. Then glue the longer bow section on to the longer hull section. The position of the mainmast will need be set forward on your new model. This is how you acquire those gigantic French and Spanish ships of 120 to 130 guns.

Unhappily, small models like this, which might prove useful for other periods, hardly seem to exist. A sixteenth-century fleet of sorts can be made from the *Revenge, Golden Hind* and *Mayflower* kits (see photograph). Neptun, the German firm, do make a few warships set in the period 1668, but these are metal and cost about 12/- each. They are scale 1-1250.

For single ship actions we *can* use the big plastic kits. These big kits are made by Airfix, Revell, Pyro and others. They are intended as front room showpieces and cost from 17/6d. to £2 each. They can be cut down to waterline models and single ship actions can be staged. The large size is unimportant because they fought very close anyway. Also they look very impressive. Once again the scale is seldom constant. If we use the Pyro *St. Louis* to fight the Airfix *Prince* we might well find that one is too small in comparison to the other. Provided that this is not *too* outrageous, it will not matter a great deal.

The German firm of Historia-Navalis manufacture scale 1-500 metal models of ships of all periods. Obviously, for 'modern' actions the scale is too large. They do have a plentiful supply of very early warships and produce 22 vessels, from the *Bremer Kogge* to the *Santa Maria*.

The latter vessel costs £1 in kit form and 50/- ready made. British taxes will up these prices by perhaps 50 per cent.

It is not terribly difficult to make your own galleons. My friend Ken Fisher has made some very nice ones using paintings from the book 'The Ship' as a guide. Sails are paper and the basic hull is simply built up in layers.

There is a further possible method of fighting battles without having to make models but using something to represent the appearance of the ships. This idea is an adaptation of the true purpose of Simplay Battle Scene series. Miniature Napoleonic ships can be stuck to an ocean background and removed at will and stuck on somewhere else. Explosions and fires can also be stuck onto the stickers.

One cannot manoeuvre the vessels for obvious reasons. A drawing of a line ship pointing in one direction cannot be made to alter its course. This problem is overcome by replacing the drawing with another that is drawn turning or heading in the desired direction. The Battle Scene sea battle, number 1 (Trafalgar), contains a great variety of English, French and Spanish ships. If a battle is fought between three or four ships, surplus drawings are used to replace vessels in the act of changing their course or drawing nearer. The idea does have limitations of course, but it *can* be made to give good results.

The stickers can also be mounted on stands and manoeuvred on the tabletop instead of on the flat ocean diorama. This is an innovation that the reader would need prepare but little effort is required. The drawings are ready made.

CHAPTER 4 **Napoleonic Rules**

Since a feature of the eighteenth century was the large number of major sea battles fought, it is important to provide rules that enable a few players to handle large numbers of warships—perhaps 20 apiece or more.

To do this, the rules must be simple and yet give us reasonably realistic results. It is all too easy to get wrapped up in a mass of detail when compiling rules. It may look impressive, it will also invariably put off the average reader straight away—for only the out and out devotee will want to see page after page of elaborate and technical rules. I hope that by presenting a set of rules that are easy to follow, you may be tempted into giving them a try.

I have included a second system of gunnery which is intended for small actions and which takes into account many factors that are ignored in the normal gunnery rules.

Whatever the situation, if a large naval battle is going to be fought, we simply cannot have complicated rules. Suppose it takes us five minutes to move each ship and undertake a series of complicated gunnery actions. It will take us *two hours* to make a single move, assuming we are commanding Nelson's fleet as at Trafalgar. The result is that the game will drag and the calculations become tedious and tactics will suffer badly. These rules will enable the game to move quickly and the players can concentrate upon manoeuvring their fleets.

When fighting smaller actions the more discerning wargamer will want to take into account the firing of each calibre cannon and shooting away enemy masts. Rules for so doing are to be found at the end of this chapter.

Since ship fighting improved but little between the years 1640 to 1840 the rules can easily be adapted to embrace the seventeenth century.

Once an initial broadside had been fired, these ships would fire off individual cannons as soon as they were reloaded. Since the quality of each gun crew varied, there would seldom be two cannons firing exactly simultaneously. However, it is certainly not practical to fire each gun individually unless we are prepared to dice 50 odd times per ship!

I have found that when cannons are fired in small batches, the result of extremities in gunfire always seem to average out by the time the whole lot have been fired. The result is that two rival warships invariably inflict almost equal damage upon each other. This can happen whatever rules are used but those contained herein make it possible to fire shattering broadsides from time to time. This is true to the real thing.

An example of what may happen is shown on the follow-

Airfix Conversions (scale: 1–600, plastic kits)
The standard *Victory* kit is in the centre. All kits have to be cut down into waterline models. The model in the foreground has been made into a two-decker seventy-four.

The other vessel has been lengthened by using two kits. Notice the join between fore and mainmasts. The cannon is a British 32-pounder made by Britains Ltd.

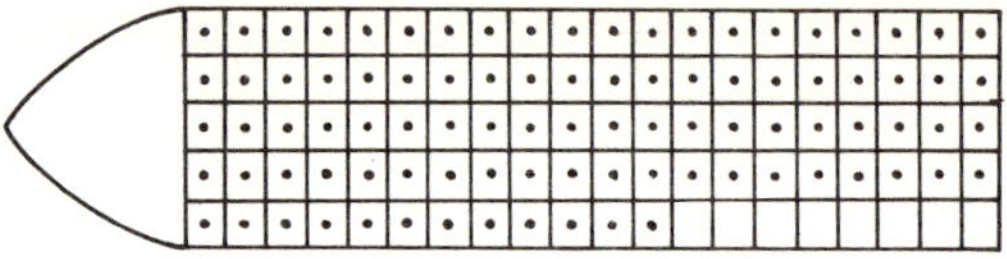

358 killed and wounded.

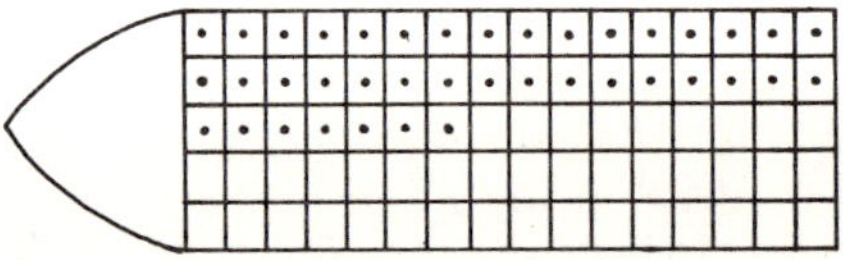

117 killed and wounded.

	LA FRANCE FIRING		*HAVOC* FIRING	
Move	*Casualties*	*Gunfire*	*Casualties*	*Gunfire*
1	29	2·9	—	2·4
2	3	2·9	9	4·7
3	—	2·9	24	2·4
4	—	2·9	4	2·4
5	4	4·6	140	28·2
6	18	3·5	4	3·8
7	—	3·5	—	1·9
8	3	3·5	15	10·6
9	17	1·7	19	1·9
10	7	3·5	—	1·9
11	—	1·2	28	5·7
12	6	1·2	19	1·9
13	28	4·6	1	1·4
14	—	1·2	—	4·2
15	—	1·2	8	4·2
16	2	1·2	85	17·1
	117		358	

ing page. The battle was a straight gunnery duel between two ships with neither side able to gain a position of tactical advantage. Notice though that the 80-gun *Havoc* was able to defeat her 100-gun opponent *La France* (without using the advantage of rate of fire) simply by more effective gunnery. Observe that of the 17 broadsides fired, two of those discharged by *Havoc* had shattering effect. The last broadside did the trick and induced the Frenchmen to strike. This crippling salvo was inflicted even though *Havoc* was herself badly damaged. The actual reason for this happening was, of course, rare and lucky dice throws.

The gunnery system is simple to use but it does involve one minor complication. You may have to find two-thirds of a figure such as 11.6 (or the nearest possible in your favour). Even those who are slow on sums should find themselves working this out in a few seconds once they have got used to the method.

Included in this chapter is a list giving all the essential details of Napoleonic warships. From this, a fleet can be made up of various rates. If it is desired to include other vessels such as Spanish or French, it is only necessary to know the same details and they can be included.

The twelfth column of this list shows the broadside value of each type and these figures are used to compute damage inflicted upon the enemy. If you introduce a type not specified on this list, you will want to calculate its broadside value. Simply halve the total weight of shot that all guns can discharge and a figure like 1,140 lbs. will be the result. Simply get rid of the last figure and alter the remaining numbers to 11.4 which is the broadside value for firing on each beam. Strictly speaking, two of the heaviest guns would only fire astern but this need not concern us, although we must account for this in a more elaborate method.

BATTLE RECORD

For each ship in our fleet we must prepare a simple squared drawing to indicate the damage she is suffering as the battle progresses.

For every gun carried, a ship is awarded one square. Thus a 110-gun ship has 110 squares and a 50-gunner is granted 50 squares. The diagram below shows how this is done. The ship is divided into 5 squared columns so that a 100-gunner will have 20 squares in each column.

Beneath the drawing we enter all her essential details and

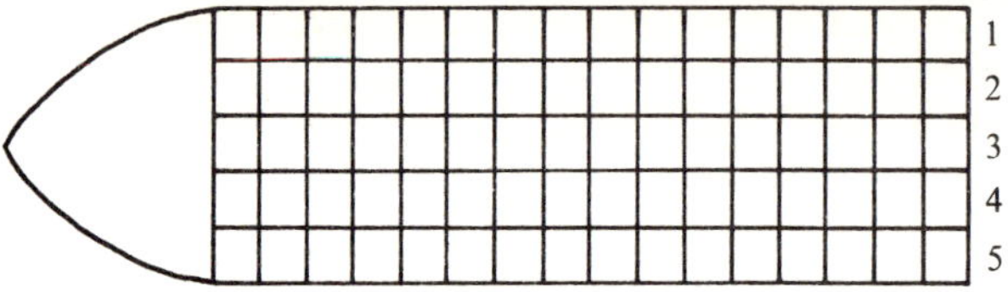

leave space to record crew casualties.

H.M.S. *Stupendous* 30-32 pdrs. B'side 9·4. Crew 719. Speed
80-gun third rate 32-24 pdrs.
14-12 pdrs.

Damage is recorded by throwing 3 dice (plus 1 red dice for crew casualties) to decide how we shall use the broadside effect figure in action. We add together the three dice scores and take note of the result on the damage inflicted chart (page 44). Let us suppose our broadside value is 8.7 and we are lucky enough to score 4 only on our three dice. This shows we multiply our broadside power by 3 and thus inflict 26.1 damage upon the enemy—truly a shattering salvo, but then the odds of scoring 4 are pretty remote.

The next step is to dot 26 squares on the ship hit by the broadside. Had the score been 26.7 we would count 27 squares destroyed because the key to the odd fraction is .7. Less and we do not count an extra square destroyed but .7 or over allows us to add an extra square to the total.

I have mentioned before that these rules are designed for speed of play and for fighting fleet actions. Therefore instead of recording demasting and number of guns dismounted, we simply know that when each column has been dotted out, there is a loss of speed and gunpower of 20 per cent. To establish the firepower remaining simply refer to the loss of firepower table located on page 44.

STRIKING

A dice is thrown when two columns have been destroyed and again thereafter each time a further column is dotted out.

Cols. destroyed	*Striking throws*
2	1
3	1.2
4	1.2.3
5	1.2.3.4

If the battle is a ship-to-ship duel, then the vessel suffering *less* damage will not strike her colours. Therefore, dice should not be cast under such circumstances.

If British or American ships strike according to these rules, the players may, if they wish, introduce saving throws (which nullifies the act of surrendering).

The Spanish were inclined to give in more readily than the French and they more so than the British.

Rate of fire

The British could fire at least twice as fast as the French and Spanish seamen and this naturally gave them a tremendous advantage. This can be taken into account by simply halving the score of the slower firing ships or by permitting them to fire only once every 3 minutes instead of once in $1\frac{1}{2}$ minutes. If it is desired to balance the odds in a set-piece battle, the French and Spanish ships could outnumber the British by 2-1. The Americans could not fire faster than the British but they were better gunners.

Elimination of gunfire

If a vessel has had all five of her squared columns destroyed without striking, then she is assumed to have been wrecked by gunfire and is no longer able to fire her guns. On the other hand, she is not sunk and can still be fired at in order to induce her crew to surrender. It is also possible to finally sink her by gunfire.

The vessel is fired at exactly as before, with a second dot added to each square as it is destroyed. When a whole column is eliminated, a dice is cast.

1. Ship destroyed by gunfire.
2.3. Crew surrender.

During firing, add crew casualties as per normal procedure.

Fires

No rules are included for vessels afire because this so seldom happened. Taking two major battles (the Nile and Trafalgar) only two vessels were burned—one in each action.

STORM DAMAGE ON CRIPPLES

Only ships crippled by gunfire are liable to be lost through storm damage. Should a storm arise, all fighting must cease and those vessels suffering damage must cast one dice for effect.

Cols. destroyed	*Dice throw*	
3	1.2	Sinks
4	1.2.3	Sinks
5 (all)	1.2.3.4	Sinks

Towing

Vessels completely demasted may be towed at one-third speed of towing vessel.

Crew casualties caused by gunfire

It has already been stated that a red dice is thrown with the three other dice that are used to determine gunnery effect. The red dice indicates crew casualties according to the number of squares destroyed by the broadside.

Red dice:		
1.2	None	(but if broadside destroys 5 squares or more, count as 1 man per square destroyed).
3	1 man	
4	2 men	
5	5 men	
6	10 men	

Casualties shown are per square destroyed.

All casualties must be recorded under the ships squared diagram.

ANGLE OF FIRE OF CANNONS

As a useful approximation, cannons have an arc of fire of 30° which is taken from the ship's mainmast. Provided this arc covers the enemies' mainmast, all guns are able to bear.

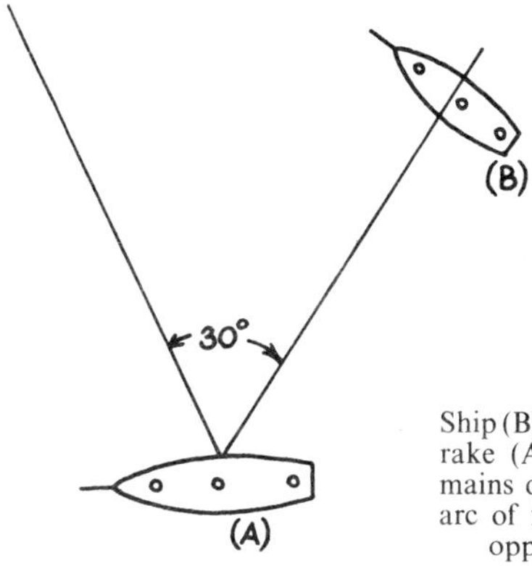

Ship (B) can easily rake (A) and remains outside the arc of fire of her opponent.

If it is desired to do this absolutely accurately, it will be apparent that more specific calculations can be made. In fact, it is possible to determine exactly which cannons can hit the enemy. The only trouble here is the time factor involved.

BOARDING RULES

When one vessel is moved so that part of her hull or bow (not bowsprit) touches a similar part of an enemy vessel, boarding is allowed. Gunnery must take place first and boarding occurs after *all* warships have fired their guns.

When warships make contact with each other (known as falling aboard) they often become entangled so that it is difficult to separate until one has struck her colours.

To break free when one-third or less of the hull is touching an enemy, a dice throw of 1 or 2 must be cast. If the contestants are alongside of each other, a throw of 1 is needed to separate.

The boarding rules are very simple but that is in keeping with the context of the whole system.

A single dice is thrown for the whole boarding party. The number on the dice is the casualties inflicted upon the enemy per 20 men in your own boarding party.

If 320 men are in action, we throw one dice. Since there are 16 twenties in the figure, we must multiply the dice score by 16. If the dice score was 5 then 16×5 are the casualties inflicted upon the enemy (80).

This is repeated until a decision is reached according to the Surrender or Repulsed table (below). A boarding move does *not* occur at every ship move. Fighting aboard is continuous until the issue is settled. The only exception to this is if another vessel is able to come alongside and add boarders within two moves.

Boarders will be repulsed and the defenders will give up the ship if either side is defeated according to the chart. 25 per cent of the casualties inflicted upon the boarders are added upon the first dice throw—but not thereafter. If both sides desire to board at the same time, a dice will be cast to decide who in fact will launch the attack first.

Defenders will *not* give up the ship if they outnumber the boarders. Boarders will *not* be repulsed if they outnumber the defenders by 2-1 or more. The chart is to be ignored under such circumstances.

If the boarders wish to retire to their own ship, this may be done without casting dice but the defenders may then follow them aboard without the 25 per cent casualty rule applying.

SURRENDER AND REPULSED TABLE

Casualties suffered	*Spanish*	*French*	*British. U.S.A.*
One-third	1.2.3	1.2	1
Half	1.2.3.4	1.2.3	1.2
Two-thirds	1.2.3.4.5	1.2.3.4	1.2.3

If one side do not surrender, or are not repulsed upon suffering one-third losses, they must cast again when further casualties have reached a half and two-thirds. If they do not stop fighting when two-third casualties have been reached, they must thereafter throw a dice after each further move to see if they give in (boarding move).

Strength of boarding parties

Normal. 25 per cent of survivors may board. If guns on one beam are abandoned, 50 per cent of survivors may board. If guns on both beams are abandoned, 75 per cent of survivors may board.

The same thing applies to the defenders. When gun crews are called up, guns may not be fired for three ship moves. One boarding move notice must be given that gun crews are to be called up. They appear the boarding move after (i.e. next dice throw for casualties). If a decision has been reached before the gun crews appear, they must surrender without fighting.

Ships do not break free of each other while boarding is taking place.

A record of losses suffered should be kept on scrap paper and when the action is over, total losses must be entered beneath the ship's squared diagram. See pages 32 and 34.

Obviously, boarding rules could be made very complicated with special rules for marines and so on. If the player wishes to use such rules, I advise that he uses this system first so that he can see just how much extra complication he can handle without causing the game to drag.

Gun range

Since the gunnery system used is simply the firing of a whole broadside a range figure must be selected at which the smaller cannons would be effective. As is well known, it was not usual for the wooden walls to fire until the enemy was quite close. A figure of 13 ins. is suggested for scale 1-1200 and double this for the larger 1-600 scale.

SAILING RULES

If it was desired to faithfully reproduce all the factors involved in manoeuvring every type of Napoleonic warship, then a truly vast set of rules would need be listed. Those given here are designed for ease of operation and to re-create the main points of sailing these vessels. In practice the results appear very satisfactory.

WIND: STRENGTH

Wind	*Green dice*	*1st-3rd rates*	*4th and 5th rates*	*6th rates*
Calm	1	X	X	X
Light	2	2	4	6
Fresh	3	6	8	10
Strong	4	8	10	12
Gale	5	12	12	12
Storm	6	X	X	X

Calm. Vessels may be kedged or towed (by small boats) at 1 knot. Storm. No fighting. Half speed in direction of wind (gale speed) or the storm may be ridden out.

Throw one red and one green dice. If the red dice registers 1, 2 or 3, then the wind strength is unchanged. Otherwise, observe green dice for new wind strength.

Wind strength and direction may change every 15 full ship moves. A throw of 1.2.3 is needed to change wind direction. Simply spin a pencil or knife. The pointed end will indicate the new wind direction.

If a gale or storm is indicated by the dice, it does not occur until 10 full ship moves later. Otherwise, the change of wind strength is within two full ship moves.

Movement

The duration of a full ship move is $1\frac{1}{2}$ minutes—the length of time it took the British to load and fire the 32-pounders.

The speed scale shown on the wind direction chart is for use with the 1-600 scale model of the *Victory* (Airfix). From these very cheap little kits, a large fleet of line-of-battle ships can quickly be acquired. The little models can be cut down to make a portion of two-deckers and otherwise altered (see photo on page 31).

The speed scale shown on the chart is for $\frac{1}{2}$-minute moves which is used when in close action with the enemy. When ships are moving but not in action, they may move a full $1\frac{1}{2}$ minutes and therefore the speeds should be trebled. Double the figures shown on the chart if you are using the scale 1-1200.

1.1/2 min.	1.1/2	1/2	1/2	1/2	1/4	1/4	1/4	1/4	1/4	1/2	1/2	1/2	1/2	1.1/2	1.1/2
		3				3					3				
			2					2				2			
			5					5				5			
					1					1					
					4					4					

As stated, when in action, a full ship move of 1½ minutes is subdivided into three ½-minute moves. Under critical conditions, if either player insists, the move may be broken down even further into ¼-minute moves. This is to prevent one player moving a ship directly in front of another so that it cannot move at all. In real life the second vessel would observe the actions of the first and alter her own course to prevent this happening.

The guns can be fired at any stage but they cannot be fired again until 1½ minutes have elapsed. In order to remember when each ship last fired, a record chart is essential. On the example above, the first line across records the duration of each move and those below register when each ship last fired (ship numbers are used).

MAXIMUM TURNING ANGLES

	64 guns plus	*32 guns plus*	*Under 32 guns*
British/U.S.A.	40°	50°	70°
French/Spanish	30°	40°	60°

These turning figures must not be exceeded and it is only possible to make one turn per move. If it is desired to make a turn, the model must be set at the desired angle prior to moving it. The ship may of course turn *less* than the figures shown.

Models are moved one at a time and in turn by each player. Firing does not occur until all have moved. If a vessel is damaged and suffers loss of firepower, it does not count until the next move. It is allowed to fire back first (unless it has already done so) before loss of firepower is counted.

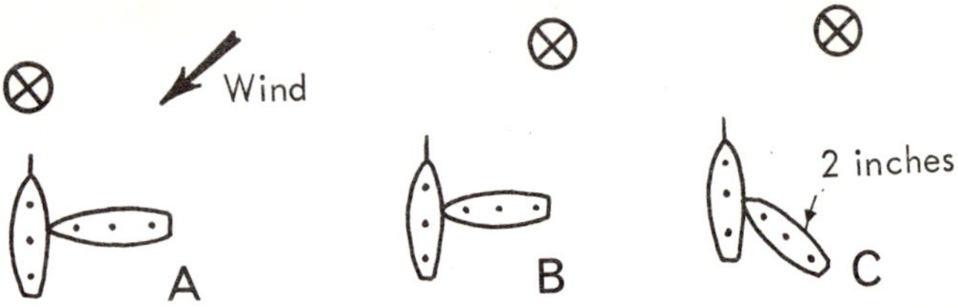

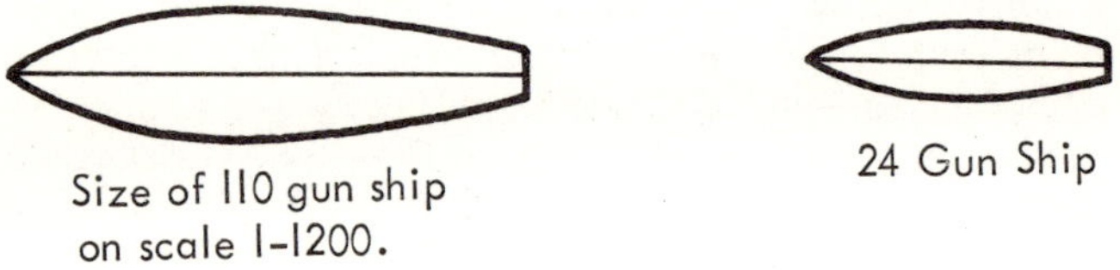

A variation of the firing system is permissible if the players wish to use it. A vessel may fire immediately upon being moved but only at another vessel that has also been moved. It is only possible to fire at one target on each beam. Gunfire can also be 'held back' and fired at any stage during the move.

Vessels interlocked in battle will drift with the wind at half speed. Ships not alongside but with bowsprit against the hull of an enemy will move as follows:

1. Find half the speed of each vessel in accordance with the direction of the wind on both.
2. Determine the average speed of the interlocked ships.
   (*a*) Ship 1 : 8 knots, half speed is 4 knots.
   (*b*) Ship 2: 4 knots, half speed is 2 knots.
   Average speed is thus 3 knots.

Note: Regarding scale 1-600, 1 knot equals 2 ins. per minute. Since combat moves are of $\frac{1}{2}$-minute duration, 3 knots equals 3 ins.

3. Move interlocked vessels in direction of the wind at average speed.
4. The stern of the vessel that has its bowsprit against the

NAPOLEONIC SHIP DETAILS

Rate	*Guns*	*Decks*	*Crews*	*42-prs.*	*32-prs.*	*24-prs.*	*18-prs.*	*12-prs.*	*9-prs.*	*6-prs.*	*B'side*	*Length*
1st	110	3	837	—	30	30	32	14	—	—	12·1	190 ft.
1st	100	3	837	28	—	28	—	40	—	—	11·6	178 ft.
1st	100	3	837	—	30	28	30	10	—	—	11·4	
2nd	98	3	738	—	28	—	30	38	—	—	9·4	
2nd	98	3	738	—	28	—	60	8	—	—	10·3	
2nd	90	3	738	—	26	—	26	36	—	—	8·6	170 ft.
3rd	80	2	719	—	30	32	—	14	—	—	9·4	165 ft.
3rd	74	2	640	—	28	30	—	—	14	—	8·7	
3rd	74	2	590	—	28	—	28	—	18	—	7·8	160 ft.
3rd	64	2	490	—	—	26	26	—	12	—	6·0	150 ft.
4th	50	2	343	—	—	22	—	22	—	6	4·1	144 ft.
5th	44	2	294	—	—	—	20	22	—	2	3·1	
5th	40	1	320	—	—	—	28	—	12	—	3·0	133 ft.
5th	38	1	284	—	—	—	28	2	8	—	3·0	
5th	36	1	264	—	—	—	26	2	8	—	2·8	
5th	32	1	254	—	—	—	26	—	—	6	2·5	
5th	32	1	215	—	—	—	—	26	—	6	1·7	
6th	28	1	195	—	—	—	—	—	24	4	1·2	
6th	24	1	155	—	—	—	—	—	22	2	1·0	113 ft.
6th	20	1	155	—	—	—	—	—	20	—	·9	

1st to 3rd rates are the line-of-battle ships.
Length is given for the purpose of model making.

enemy hull, must be swung round 2 ins. (1 in. if scale 1-1200). This should be done in relation to wind direction. This is repeated each ½ minute move.

LOSS OF FIREPOWER TABLE (LARGE SHIPS)

Ship	*1 col.*	*2 cols.*	*3 cols.*	*4 cols.*	*Full*	*No. of squares in each column*
110	9·7	7·3	4·9	2·5	12·1	22
100	9·2	6·9	4·6	2·3	11·6	20
100	9·2	6·9	4·6	2·3	11·4	20
98	7·6	5·7	3·8	1·9	9·4	3 at 20. 2 at 19
98	8·3	6·2	4·2	2·1	10·3	3 at 20. 2 at 19
90	6·8	5·1	3·4	1·7	8·6	18
80	7·6	5·7	3·8	1·9	9·4	16
74	6·9	5·2	3·5	1·7	8·7	4 at 15. 1 at 14
74	6·2	4·6	3·1	1·6	7·8	4 at 15. 1 at 14
64	4·8	3·6	2·4	1·2	6·0	4 at 13. 1 at 12
50	3·3	2·5	1·6	·8	4·1	10

This table indicates firepower remaining after each squared column has been destroyed (dotted out on damage control drawing) and column 7 shows the number of squares in each of the 5 columns.

To compute firepower remaining on other ships, simply deduct 20 per cent of the total firepower per column destroyed.

DAMAGE INFLICTED CHART

Dice throw		*Dice throw*	
9.10.11.12	¼ only	5.16	×2
8.13	½ only	4.17	×3
7.14	¾ only	3.18	×4
6.15	Normal		

If the broadside value of your ship were 12.1 and a score of 7 were cast, ¾ of the broadside is counted (9 squares destroyed)

GUNNERY FOR SMALL ACTIONS

For the purposes of recording damage, only crew casualties are taken, but here we must see the casualty figures as also representing the battering the vessel is taking. After all, if 100 men are killed by cannon fire, a fair amount of damage must have occurred.

The bigger and more powerful the vessel, then the greater number of crew that is carried. The Spanish did tend to overman some of their huge men-of-war, so a portion may be deducted if this is evident.

Striking colours

A player may refuse to strike if escape is possible or if rescue is at hand. Otherwise he *must* surrender in accordance with this table.

Casualties suffered	*Spanish*	*French*	*British/U.S.A.*
One-third	1 to 3	1 or 2	1
Half	1 to 4	1 to 3	1 or 2
Two-thirds	1 to 5	1 to 4	1 to 3

(Single dice throw)

Part of the wargames fleet of Ken Fisher and the author. Note the big 120-gunner second from the rear of the Spanish line. Also present are 100-gunners, seventy-fours, a sixty-four and a fifty. There is also a scratch-built frigate. An attempt is being made to break the Spanish line. These small Airfix kits cost less than 15p each.

When four-fifths of a crew are casualties, surrender by any nation is compulsory unless rescue is immediately at hand.

It is obviously not necessary to prepare a squared diagram of each ship as was required in the simpler set of gunnery rules. However, a fleet list is still required. Each vessel should be named and numbered and the details of the ship should be entered under her name with sufficient space to record damage.

FIRING TABLE

Range	*42 pdrs.*	*32 pdrs.*	*24 pdrs.*	*18 pdrs.*	*12 pdrs.*	*9 pdrs.*	*6 pdrs.*
0-2	4·2	3·2	2·4	1·8	1·2	·9	·6
3-4	4·0	3·0	2·2	1·6	1·0	·7	·5
5-7	3·8	2·8	2·0	1·4	·8	·5	·3
8-10	3·6	2·6	1·8	1·2	·7	·4	·2
11-13	3·4	2·4	1·6	1·0	·6	·3	·1
14-16	3·2	2·2	1·4	·8	·4	·2	
17-19	3·0	2·0	1·2	·6	·3	·1	
20-22	2·8	1·8	1·0	·5	·2		
23-25	2·6	1·6	·8	·3	·1		
26-28	2·4	1·4	·6	·2			
29-31	2·2	1·2	·4	·1			
32-34	2·0	1·0	·2				
35-37	1·8	·8	·1				
38-40	1·6	·6					
41-43	1·4	·4					
44-46	1·2	·3					
47-49	1·0	·2					
50-52	·8	·1					
53-55	·6						
56-58	·4						
59-61	·2						
62-64	·1						

The range figures are in inches and are for use with scale 1-1200 models, (60 ins. equals 1 mile). If 1-600 models are being used, simply double the range figures shown here. At Trafalgar the French opened fire at 1,000 yds. The 42-pounder could throw a ball about 2 miles but at such range it would be non-effective.

The figures represent the average firepower of each calibre cannon at the range given. To compute the effect of a broadside of guns able to fire, let us assume we are firing 14 24-pounders. Two have been dismounted and only two-thirds of those remaining are loaded (further rules explain this). Thus 8 cannons are able to fire. The range is 7 ins. Thus the figure given is 2.0 per cannon. (see above) 8×2.0 is 16.

The firepower of the 8 cannons is thus 16 but now we must dice to see just how effectively we fire them.

Dice	*U.S.A.*	*British*	*French*	*Spanish*
1	$\frac{1}{2}$	$\frac{1}{4}$	$\frac{1}{8}$	—
2	$\frac{3}{4}$	$\frac{1}{2}$	$\frac{1}{4}$	$\frac{1}{8}$
3	Norm	$\frac{3}{4}$	$\frac{1}{2}$	$\frac{1}{4}$
4	Norm	Norm	$\frac{3}{4}$	$\frac{1}{2}$
5	$\times 2$	$\times 2$	Norm	$\frac{3}{4}$
6	$\times 4$	$\times 4$	$\times 2$	Norm

Thus if the British 8 24-pounders register a dice throw of 3, then three-quarters of the firepower figure of 16 is taken (12 casualties). Had a 6 been diced, then the 16 would have been multiplied by 4 and caused 64 casualties.

It will be apparent that extreme results can occur-from casulaties of only a few to two or three hundred (full broadside at point blank). These extremes will be rare but a feature of the fighting in this period, is that broadsides could sometimes be absolutely shattering and almost non effective on other occasions (i.e. if the guns were fired late on the downward roll).

All pieces of different calibre are fired separately. Fire 32-pounders first, followed by the 24-pounders and then the 18-pounders, etc.

Firing full broadsides

Full broadsides may be fired (by guns in range) provided that no cannons have been in action for the past two moves. In other words, three minutes must elapse before full broadsides can be fired without having to cast dice to see how many cannons are loaded.

Throw	*1*	*2*	*3*	*4*	*5*	*6*
42 pdrs.	$\frac{1}{3}$	$\frac{1}{3}$	$\frac{1}{2}$	$\frac{2}{3}$	$\frac{2}{3}$	All
32 pdrs.	$\frac{1}{3}$	$\frac{1}{2}$	$\frac{2}{3}$	$\frac{2}{3}$	All	All
24 pdrs.	$\frac{1}{3}$	$\frac{1}{2}$	$\frac{1}{2}$	$\frac{2}{3}$	All	All
18 pdrs.	$\frac{1}{2}$	$\frac{2}{3}$	$\frac{2}{3}$	All	All	All
12 pdrs.	$\frac{1}{2}$	$\frac{2}{3}$	All	All	All	All
9 pdrs. and 6 pdrs. All						

French: Count number thrown as 2 less unless low number thrown.

Spanish: Count number thrown as 3 less unless low number thrown.

I.e. if Spanish threw No. 5, count number 2. If 1, 2, 3 or 4 thrown, count number 1.

Dismounted cannons

Note total hull damage scored at end of ship firing. Throw 1 dice per 20 casualties scored. The number of cannons dismounted per 20 men killed or wounded is as shown on the table below.

1	*2*	*3*	*4*	*5*	*6*
None	None	1 gun	2 guns	3 guns	4 guns

The number of cannons dismounted must be noted under the heading of the ship concerned in the fleet list. Deciding which calibre guns are destroyed must be done by sensible reasoning of the numbers of each calibre cannon carried.

EXAMPLE

9 cannons	*Guns carried*	*Port*		*Starboard*	*Astern*
have been	30-32 pdrs.	14	(2)	14	2
dismounted on	30-24 pdrs.	15	(3)	15	
the port side	32-18 pdrs.	16	(3)	16	
of the ship.	14-12 pdrs.	7	(1)	7	
			9 cannons		

Figures in brackets are the cannons dismounted. Note that it is impossible to 'distribute' cannon losses with complete accuracy but the manner in which this has been done above is as fair as possible.

As far as the example goes, port broadside is reduced to 12 32-pounders, 12 24 pounders, 13 18-pounders and 6 12-pounders, next time the ship fires her guns. Further, they may not all be loaded and able to fire. This will depend upon the dice throw relating to numbers of cannons reloaded per 1½ minutes.

FIRING AT MASTS

It is not possible to fire at masts when vessels are closely engaged. The principle is the same as when firing at the hull, with the exception that the score is to be regarded in terms of points value and *not* crew casualties.

The score obtained is used to determine the number of dice that may be cast. Each dice throw of 1 or 2 permits damage to masts and then the next table is used to register the damage caused.

Dice throws for mast damage according to score

Score	*Dice throws*	*Score*	*Dice throws*
10-20	One	101-150	Four
21-50	Two	151-200	Five
51-100	Three	Over 200	Six

If a player states that he is firing bar or chain shot, the points score may be added by 50 per cent. However, that portion of bar and chain shot which misses masts and strikes the hull, will cause only one-third of normal damage.

MAST DAMAGE

Dice Throw

1. Mizzen topmast or Bowsprit (dice for which).
2. Mizzen Mast.
3. Fore topmast.
4. Foremast.
5. Main topmast.
6. Mainmast.

The ideal effect of demasting is to reduce the speed of the ship by a percentage according to which mast is shot away. In practice, this leads to enormous complications.

Instead, the ship is only reduced in speed when two whole masts have been brought down. Her speed is then reduced by half. Otherwise, the effect is to lose a specified number of $\frac{1}{2}$-minute moves.

Mast shot away	*Loss of short moves*
Mizzen topmast/Bowsprit	One
Mizzen mast	Three
Fore topmast	One
Foremast	Three
Main topmast	Two
Mainmast	Four

If all masts are brought down the ship cannot move at all. They can be towed however (see page 40). Also temporary repairs can be made enabling the vessel to sail at one-third speed. Repairs take 15 minutes or 10 full ship moves and cannot occur if vessel is under fire.

By marking the movement record (page 41) an easy check can be kept on when each ship is permitted to move again.

Which masts are lost should also be recorded in your fleet list.

OPTIONAL RULES

Shots missing masts and hitting hull

Cast one dice per 20 scored and each throw of 1 or 2 will result in a quarter of the total damage to masts being converted into crew casualties.

Shots missing hull and hitting masts

For every 20 casualties scored by cannon fire, throw one dice. Each 1 cast entitles a separate throw to see which mast has been shot away. Use table above. Although not possible to *aim* at masts when closely engaged, they can still be brought down by use of this method (masts shot through below decks).

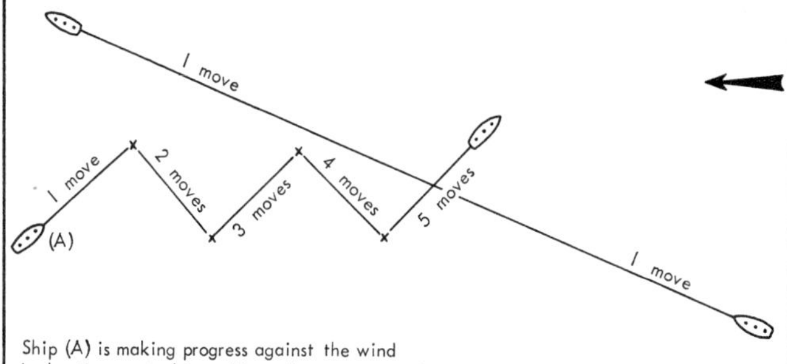

Ship (A) is making progress against the wind by beating or tacking. Note distance covered compared with other ship.

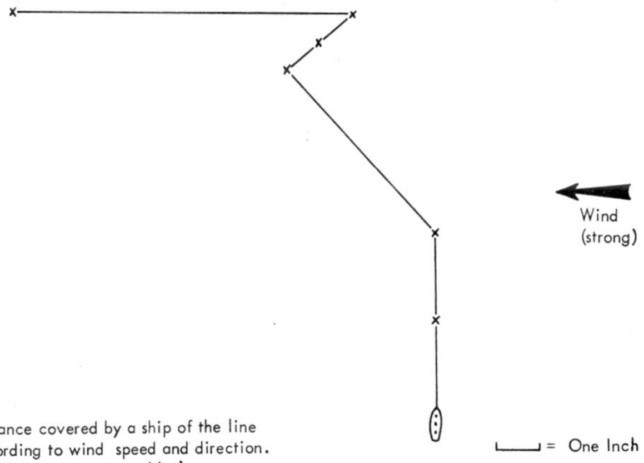

Distance covered by a ship of the line according to wind speed and direction. (See wind direction table)

SEQUENCE OF EVENTS

Preparing the Fleet list

Having selected the vessels that will take part in the campaign or setpiece, each should now be entered in a fleet list as indicated in the rules. Ships are entered in order of size or number of guns and are numbered from (1) up to the total number of vessels. Beneath the name of each ship are recorded full details of her particulars with space left to record damage. When the simplified version is being used, a squared diagram of each vessel is necessary. See pages 32 and 34.

Movement

(1) Cast dice for wind speed and direction on every 15th move. Place the wind direction table on the battle area so that vessel movement can be calculated at a glance.

Move models in time durations of $1\frac{1}{2}$-minute moves until vessels are in action. Thereafter movement is reduced to $\frac{1}{2}$-minute moves until vessels are in close action when $\frac{1}{4}$-minute moves may be taken. Move models unable to engage first (i.e. out of range) and then move vessels in battle. When warships open fire, record same on record chart (page 41).

GUNNERY

Since the first set of rules are a simplified version, only the sequence of firing for the second more elaborate set of rules are presented here.

1. Determine which ships are able to fire upon the current move. Consult record chart.
2. Measure range to target.
3. Calculate whether guns can bear upon target.
4. Consult fleet list to see how many and which cannons have been dismounted by enemy fire.
5. Throw dice to decide how many of the cannons remaining are reloaded in time to fire (see page 48). This is not

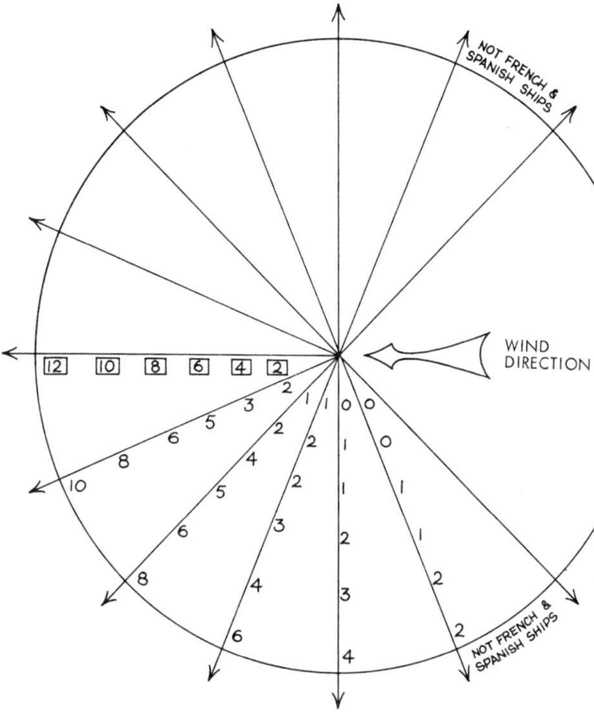

Wind Direction Table

The purpose of the above diagram is to enable players to move their models with a minimum of calculation but retain a fair degree of accuracy. Only an immensely complicated system could faithfully reproduce the exact manner in which these wooden walls had to be sailed, for to do so was both a science and an art.

Trace out a copy of the diagram and place it in position on the floor or table with the arrow pointing in the direction of the prevailing wind.

The key figures are those that are boxed. Simply find the speed of your ship (which varies according to wind strength) and by simply noting the direction that she is sailing the relevant speed can be seen.

If a vessel has been damaged and (say) reduced to half speed, she must sail at HALF the speed shown on the diagram.

It is possible to make progress against the wind by beating or tacking (see diagram), although this will be a painfully slow process.

necessary if the guns have not been fired for three minutes.

6. State whether target is masts or hull each time a set of different calibre cannons are fired. Also state whether firing ball or bar and chain shot.
7. Cast dice for effect of gunfire as instructed on pages 46 and 47 for hull damage and page 49 for mast damage.
8. Record damage caused in the fleet list and throw dice to see how many cannons are dismounted by this damage.
9. After every vessel has moved, determine where boarding is possible and fight boarding actions.

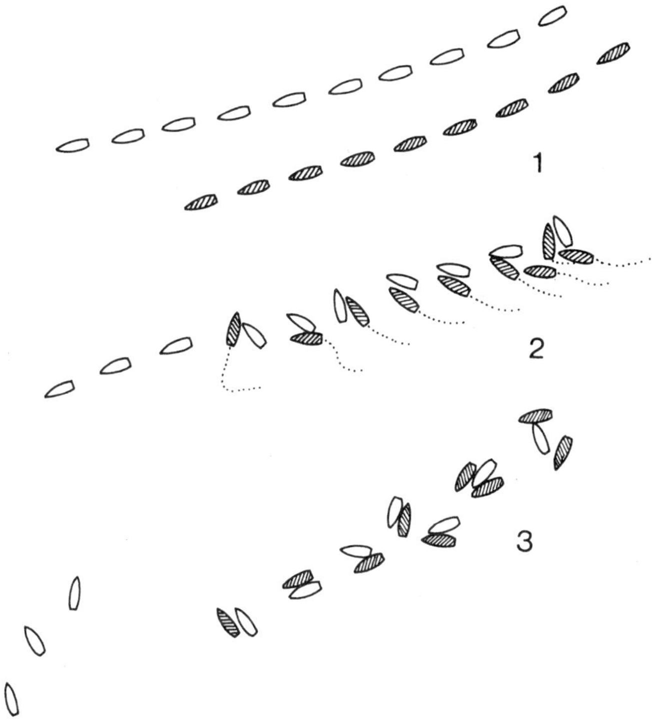

Napoleonic Tactics

(I) Typical gun battle which seldom has decisive results.

(2) On some occasions, an attempt was made to break the enemy line and isolate the van.

(3) By the time the van has returned to action, the rear is overwhelmed and the centre demoralised. Such tactics also bring about a melee which is avoided by the normal lines of battle as in (I)

In practise a break-through is difficult to achieve.

CHAPTER 5

Ironclad Period Rules for the Period 1860–70

The original intention was to dedicate this chapter to the reconstruction of the Battle of Lissa 1866. However, this would have involved scratch building two very large foreign fleets and may have proved a task too difficult for the average reader. Instead, I have simply written rules for fighting set-piece battles set in the period. The wargamer is thus free to stage whatever type of action he chooses, be it a single ship affair or something grander. He can also decide which nations' ships he will command. One big advantage of fighting in the 1860s and '70s is that the warships are very easy to model. There is no superstructure to speak of and the hulls were usually very uncomplicated.

The gunnery chart has been based upon the performance of the Austrian and Italian guns. If the reader chooses to include gun calibres not shown here, it will not be difficult to include them. Once the principle of the chart is understood you will be able to make a chart similar to that given here but including other types of guns. Some older types of ship mounted guns fired 15-in. steel balls! All that is needed is the information on what the gun could do in general and the weight of shot and range and this can be built into the chart by marking down the appropriate figures in the correct column.

About the time 1866, warships were undergoing a remarkable transition and so were the guns they carried. At Lissa, the hardcore of both fleets consisted of the ironclads but wooden ships there were in plenty. These proved vulnerable to shell fire whereas the ironclad was almost immune. Needless to say, the wooden warships played an important part in the battle. All vessels were driven by steam but with one exception (the ram-ship *Affondatore*) were all sail assisted. The wooden two-decker *Kaiser* which played an heroic part in the famous action, was in basic layout very little different than the ships that fought at Trafalgar. Guns were mostly carried broadside fashion but it is interesting to see the advent of the turret mounted aboard one of the Italian warships.

The same link with the past, present, and future is to be found in the great variety of guns that thundered forth off the island of Lissa over 100 years ago. There were rifled guns and smooth-bores. Some were muzzle loaded and others had breeches. Nor had the shot escaped the transitional difficulties. Explosive shells were used but so also was solid elongated shot and steel cannon balls.

Let us now see how effective the various types of naval gun were.

1. *The rifled breech loaders*

   At Lissa the Austrians do not appear to have been much troubled by them but the British found them a costly failure. Average life of one of these guns was only 13 rounds and the shooting was erratic.

2. *The rifled muzzled loaders*

   Slow loading but accurate at fairly long range. Since velocity of the shell was well maintained throughout its flight the projectile could penetrate at distances greatly beyond that of the smoothbore. The gun could also outrange the faster firing smoothbore.

3. *Smoothbore muzzle loader*

   Rapidity of fire combined with greater penetration at close range made this gun the most popular in the Royal Navy. Up to 1,500 yards or so the smoothbore was as accurate as the rifled gun.

All these factors I have attempted to account for in the gunnery tables and charts. The fact is though, that only rarely will shots cause great damage to Ironclads. Most hits will simply strike the hull and cause minor damage only. The smoothbores mostly fired solid shot which could penetrate armour better than shell fire. It also had a more battering effect than shells even when failing to pierce a ship's armour.

I have not sought to include rules for disabling individual guns by individual shots because of the extra complication this entails.

Should players wish to include this factor, remember that only rarely would a projectile enter an open gun port. It does not appear to have happened at Lissa.

The main purpose of gunfire is to disable an opponent so that the ram can be used with deadly effect. Decisive results *can* be obtained against Ironclads by gunfire—but only rarely. The iron sided ship of the 1866 period had a notable Achilles heel in the lack of armour to prevent shots entering at the stern or bow. The Italian *Palestro* was destroyed in this fashion by a shell that caused a fire which reached ammunition.

Great faith was placed in the ram but it must be stressed that it proved all too easy to avoid an aimed blow by an enemy ram. Even if the blow could not be avoided altogether, it could be 'caught' by deft use of the helm so that only a glancing strike was achieved. All this was true when a vessel had seaway. When she was boxed in or disabled, it was a different story. A square blow will invariably cause fatal damage.

PREPARING THE FLEET LIST

The points system for evaluating the warship of the period is simplicity itself.

Points value

I point per ton. If vessel is Ironclad, award 2 points per ton. Despite its apparent crudity, this simple method is as satisfactory as working out some complicated formula. It is a general principle that apart from how well built she is, a vessel is more difficult to sink than another that is smaller.

Since there is no difficulty in calculating the points value of any warship, let us pass on to the damage record.

Each ship is entered on a sheet of paper or on a page in an exercise book if a lot of vessels are involved. Enter ships in sequence of size and importance so that the largest appears at the front and the smallest at the rear. Put each ship's name at the head of a page and enter beneath her name all the essential details.

EXAMPLE

<table>
<tr><td>IRONCLADS "KAISER MAX". 11 Kts. 7,176 P.V.
Crippled at 3,588 P.V.</td><td>GUNS 14-64 pdr shell guns
16-55 pdr S.B.M.</td></tr>
<tr><td rowspan="4">POINTS LOST
24
7
140
10
181
14</td><td>SPECIAL DAMAGE
Boiler hit.</td></tr>
<tr><td>MASTS 1. Shot away</td></tr>
<tr><td>SAILS 1.
2.</td></tr>
<tr><td>FIRES</td></tr>
</table>

During the fighting, the damage inflicted and recieved is jotted down on scrap paper as it happens. The total damage inflicted by one ship upon another is entered immediately after she has fired off all her guns. Any damage suffered is added to the points already lost. When special damage is suffered, it is entered in the appropriate column but in the case of fires, must be crossed out if the fire is extinguished.

DAMAGE

Loss of speed through demasting and hits in vital quarters is dealt with under the Special Effect Hits section. Loss of control means that a vessel must proceed on present course, or circle according to a dice throw. Deliberate manoeuvres are not allowed.

1.2. Proceed on course.
3.4. Circle to port.
5.6. Circle to starboard.

There is no provision for the destruction of individual guns. Instead we simplify matters by stating that when half of the vessels points are lost, then also we reduce by half the number of guns that may be fired. Speed too is halved.

Obviously, this can be taken further. For example, when 20 per cent of a ship's points value is lost, so is 20 per cent of her firepower and speed. Do not include such rulings until you are well practised at the game, because it will add greatly to the amount of work and checking to be done.

A vessel that is appropriately marked will serve as a reminder that she is crippled and may discharge only half her broadside and steam at half speed. A small piece of cotton wool, with its resemblance to a fire aboard is a suitable method.

Ramming

The damage inflicted by a vessel upon another when striking it, depends upon the following three factors.

1. The angle of the blow.
2. Size of vessel ramming.
3. Speed of ramming vessel.

Angles of ramming

(A) 60 to 90 degrees. (B) 40 to 59 degrees. (C) 20 to 39 degrees.

Damage inflicted by ramming

Angle	*Ironclad*	*Wooden*
(A)	1,000 pts per kt, plus p.v. remaining of ramming ship:	ditto but 500 pts per knot.
(B)	Half of remaining p.v. plus 500 pts per kt.	ditto but 250 pts.
(C)	One quarter of remaining p.v. plus 200 pts per kt:	ditto but 100 pts.

Note: If vessel is less than 2,000 tons, reduce final figure by one-third.

Whenever ramming occurs and after damage to rammed vessel has been computed, a dice is thrown to see if ramming vessel also suffers damage.

Angle	*Ironclad*	*Wooden*
(A)	1	1.2.3
(B)	1.2	1.2.3.4
(C)	1.2.3	1.2.3.4.5

If any dice throws as indicated above are thrown, then another dice is cast to see the amount of damage suffered by ramming ship.

Angle	*Ironclad*	*Wooden*
(A)	2%	3%
(B)	4%	5%
(C)	8%	10%

The percentage figures given are per spot on dice throw. i.e. Ironclad at angle (B) throws No. 4. Four per cent damage per spot equals 4×4 per cent or 16 per cent. Damage is percentage of points value remaining of ship ramming. For those not so good at sums, an easy way to calculate percentage damage is as follows. What is 8 per cent of 3,700? Deduct 80 from every 1,000 and 8 from every 100. Damage thus 294.

It is quite in order for wooden vessels to ram ironclads which had no outstanding advantage under such conditions. The armour plating was easily displaced. The armoured ships were able to resist ramming a little better than the wooden ships and this is accounted for by their extra points value.

Ramming procedure

At angle (C) the vessels are not interlocked but will scrape alongside each other still moving at the correct speed but altering direction when contact is made, (models swung alongside each other). This type of collision is liable to bring down a mast or two. Throw one dice per mast and each throw of 6 denotes a mast down.

At angles (B) and (A) the bow of the ramming ship is driven into the hull of her victim. The vessel may not withdraw until the following short move and must then back off at one quarter normal speed.

What happens to the motion of the ships when the collision occurs? In fact, it should prove almost impossible to hit another vessel a square blow when she is moving at a fair speed. Only those boxed in or disabled could be rammed properly. If however, this situation should arise, then simply leave the models in the position they are at the point of contact. If the rammed ship is not sunk, it may pick up speed after the other vessel has withdrawn its bow.

In practice, vessels should not pick up more than 2 knots speed per short move (15 seconds) but it will prove very difficult to remember at what speed each vessel was moving on the previous move. However, this ruling can safely be applied to warships that have engaged in ramming. It will also be apparent that ships cannot simply come to a dead stop (i.e. to avoid collision with a consort). It is therefore not possible to reduce speed by more than 2 knots per short (ramming move) or 8 knots in a full one minute move.

Movement Rules

It is impractical to use a scale much larger than the popular 1-1200 and the scale given here is written for models of this size. A large vessel will be about 3 ins. long so that a big table or a suitable floor space will be adequate in which to conduct the action.

The scaled speed of any 1-1200 model is 1 in. per knot per minute, so that a 9 knot vessel will move 9 ins. in each minute.

Scale examples

1 in. = 33 yds.	15 ins. = 500 yds.
3 ins. = 100 yds.	60 ins. = 2,000 yds. (1 nautical mile)
6 ins. = 200 yds. (one cable)	

If the wargamer selects a scale other than 1-1200, he will need alter the figures given here.

When models are in close proximity of enemy warships,

it is always necessary to alter (reduce) the duration of the move. The reason for this is that if both sides make lengthy moves, then it is not possible to counter the tactics of the enemy. For example, let us imagine two model fleets steaming in column and heading in the same direction. Now (A) moves first and continues on course with his leading model. (B) however, uses his turn to make a drastic alteration of course directly towards (A) with his leading ship.

The player (A) may not wish to fight a pitched battle but he now must either abandon the leading ship or support it and accept a pitched battle. Although he can move away with the remaining ships, he has committed the leading vessel. In real life, the move would be countered by all vessels as it happened.

That is but one situation. There are very many other unrealistic things that can happen if each player in turn takes a long move. It is even worse if we allow one side to move all his vessels first. Each model *must* be moved in turn. Players must move in rotation—one ship at a time.

The duration of the moves are as follows.

(a) 1 minute full move (12 ins.)
(b) 30 second combat move (6 ins.)
(c) 15 second ramming move (3 ins.)

Figures in bracket equal distance moved by 12 knot vessel.

The 30 second combat move is used when rival warships close to within 45 ins. of each other (1,500 yards) so that sudden moves by the enemy may be more reasonably countered.

The ramming move of 15 seconds is used only by rival models within 12 ins. (400 yards) of each other. No ramming may be attempted unless the models involved are using the very short move. Such vessels make two moves for every vessel not endangered by ramming.

Players move a ship each in turn. Sequence of events during moving is as follows:

1. Move vessels well out of the battle first. These can be

moved for one minute but must not be moved again until other ships in action have moved two combat moves or four ramming moves.

2. Move models engaged in gunnery but in no danger of being rammed (30 second move).
3. Move the models that are striving to ram or avoid the ram. Each model is moved twice at $\frac{1}{4}$ speed but strictly in rotation.

Turning rules

Since time and distance is related to the size of the model, so must be the turning circle of the ships. Unhappily, the turning circle varies somewhat from ship to ship.

The Italian ram-ship *Affondatore* was a very unhandy ship that required 3,600 yards ($93\frac{1}{2}$ ins.) to describe a complete circle at about 11 knots. The best figure for an Ironclad appears to be 560 yards or 17 ins. but an *average* would appear to be about 850 yards or 25 ins.

This average turning circle is shown below and when it is desired to make a maximum turn, it will be necessary to use the curved cut out.

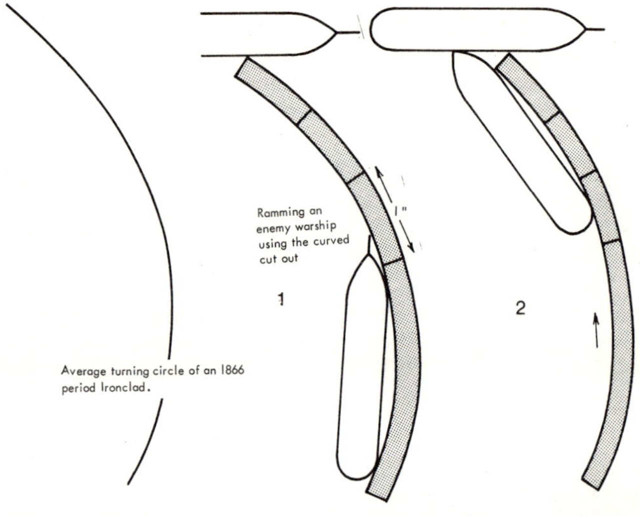

Three cut-out's are ideal. One will have 6 ins. of movement (plus a couple of inches to position the model) the second 3 ins. and the third one inch. See the photograph on page 66 illustrating these three turn indicators. They are easily cut from balsa wood.

Making the turn indicators (cut-out's)

1. Cut a length of string 25 ins. long.
2. Select a saucepan lid (or similar round object) that the string will circle without being more than an inch or two too long or too short.
3. Use the lid to draw a circle upon a section of balsa or other soft wood. The whole of the circle is not required but only sufficient to make the three turn indicators described above. Allow 2-3 ins. of the curve to position the model and scale off the rest in inches (scale off the string in inches to measure the curve accurately).
4. Cut out the indicators with a sharp knife. The curve used to turn the models must be the *inside* curve, so cut between $\frac{1}{4}$ and $\frac{1}{2}$ in. away from it so that the whole turn indicator is that width.

We shall therefore have made three turn indicators of different lengths. The shortest is used when there is simply not room (among the models) to lay out the longer ones. The largest indicator can only be used when conditions permit it to be laid down without fouling another model. Under normal conditions the 3-in. cut-out will be used when attempting to ram. In confined spaces the 1-in. indicator can be used (note the handle fitted so that it can be lowered from above into a congested area) but will only permit the model to advance 1 in. at a time.

The model should have the cut-out fitted snugly against it with the bow and stern touched by the inside curve. The ship is positioned so that the bow is exactly level with an inch calibration mark. It can then be moved forward the desired number of inches (knots).

There will be a small gap where the curve of the cut out sweeps away from the main hull. As long as the cut-out is

(1) Using the curved cut-outs or turn indicators. See text. Some later period models have been used to represent ironclads. H.M.S. *Nile* is attempting to ram the German *Preussen*.

(2) A turn indicator relevant to the space available is placed in position. Notice that H.M.S. *Nile* can only ram the enemy's bow. The indicator is scaled off in inches.

placed correctly to touch the bow and stern, all will be well. Hold the turn indicator down firmly with one finger and slide the model along the curve so that it follows exactly the maximum permitted turn. When the model can reach its target, place the turn indicator so that the far end lightly touches the victim. When the model is slid along to the end, she will be touching her enemy in the ramming position. Obviously, if it is desired to position the model at an angle *less* acute than that dictated by the turn angle, then this may be done.

The angle of the turn is almost exactly equal to a series of turns made at 35° every 3 ins. (knots) in *straight* lines. When conditions dictate, moving in this fashion is permitted. A thin slip of paper made in the fashion of the sample shown on page 66 can be slipped under a model and the maximum angle at which she may turn is at once displayed.

Great care must be taken not to disturb models prior to moving them. If a model ship is accidently moved, it may

(3) A finger is placed on the turn indicator to hold it steady and the model slid along it until the *Preussen* is touched.

Damage is computed according to speeds, size of ramming ship and angle of blow.

Mercator models. This German firm makes beautiful little models of early warships. These are a few of them. Historic passenger ships are also manufactured. The ironclad turret ship is 3 inches long.

have vital consequences or lead to a bitter dispute between the players.

For the most part, it will not be desired or required to make extreme turns. The cut-out turn indicators are mostly used under critical conditions when a maximum turn *is* needed (i.e. to ram or avoid the ram, or to change direction as quickly as possible).

Otherwise, if a ship is obviously not going to exceed the maximum permitted turning circle, the model may be moved with a straight edge rule.

Using these rules the players will find that it is always possible to avoid a direct blow by an opponent *unless* the models are handled badly in a tactical sense. On the other hand, once the ships melee, a collision of some sort is difficult to avoid altogether. Ramming bow to bow should not be allowed although it is possible in practice.

Boarding

It is possible to make boarding rules for this period but because it happened so very rarely, I feel it best not to include them.

Rate of fire: Column 7 of the *Gunnery Information Chart* (A) shows the approximate rate of fire of each type.

Now it is strictly an optional feature to use these rates of fire. The 'Hard & Fast' rules are listed below, because although they are not as realistic, they are much easier to use. It is felt that after some experience, the players may like to try their hand at this optional rule. Note however, what is involved.

The wargamer can fire each set of different guns carried aboard his ship at any stage of the game. For example, he may fire off his smooth-bores and withhold discharge of his Rifled pieces. Also he can confine his gunfire to only a

GUN	SHOT	NON PEN DAMAGE	LONG RANGE	MEDIUM RANGE	CLOSE RANGE	RATE OF FIRE
10 inch	295 lb. Shell	12	90"-51"	50"-36"	35"-20"	4 Mins
8 inch Rifled	147 lb. Shell	5	80"-46"	45"-31"	30"-15"	3 Mins
8 inch S.B.M.	55 lb. ball	7	70"-46"	45"-31"	30"-15"	2 Mins
7 inch Rifled	74 lb. Shell or 107 lb. bolt	4 10	80"-46"	45"-31"	30"-15"	3 Mins
7 inch S.B.M.	40 lb. ball	6	70"-46"	45"-31"	30"-15"	2 Mins
6.3 inch Rifled	64 lb. Shell	3	78"-46"	45"-31"	30"-15"	3 Mins
60 pdr. Rifled	47½ lb. Shell	2	74"-44"	43"-29"	28"-15"	2 Mins
30 pdr. S.B.M.	32½ lb. Shell or 24½ lb. Ball	1 2	62"-44"	43"-29"	28"-15"	1 Min
4.7 inch Rifled	25 lb. Shell	1	70"-46"	45"-31"	30"-15"	2 Mins
12 pdr*	12 lb. Shell	½	62"-44"	43"-29"	28"-15"	30 Secs

* CAN ONLY PENETRATE IRONCLADS AT CLOSE RANGE

CHART (A)

portion of the ships in range-perhaps wishing to withhold fire of the others until a better chance of hitting is presented.

Obviously a record of the time the guns were fired must be kept so that they cannot be fired again until the number of moves corresponding to the rate of fire of the piece has elapsed. A practised wargamer and a practical method of recording time and the firing of guns will not make this too difficult.

The normal method of gunnery is described as follows:

The Guns are fired every 3 minutes of game time. It is not possible to fire any guns other than at this time which may be described as the *Special Gunnery Move* (S.G.M.). Even if a warship is but 1 in. out of range, it may not fire until the next S.G.M.

Every S.G.M. the Rifled guns are fired once and the Smooth-bores twice. Obviously, this only applies to guns in range of an enemy ship.

Exception: the solitary exception to this rule concerns the 10-in. guns of the Italian Ram-ship *Affondatore* which can only be fired once every *4* minutes.

Note: A non-operating alarm clock is most useful for recording time. The minute hand is put forward one minute at a time by the players. Otherwise, register the passage of time on a special sheet.

METHOD

A complete gunnery sequence of events is listed (see page 72) which is designed to explain step by step the procedure from the start of a gunnery move to the finish. Those points covered are not repeated here, except when it is necessary to elaborate on a point. That which is covered here is more in the nature of rules concerning the sequence of events.

(a) It is not possible to fire at more than two targets unless certain guns can *only* bear on a third target.

(b) To avoid unnecessary complication, it is assumed that half the total guns carried by each warship can fire on either beam, and that two of the heaviest guns can fire ahead and two astern. In practice, this ruling is absolutely correct in about 80 per cent of instances.

(c) If a ship is crippled by gunfire (see page 60) the loss of a portion of its gunfire is not counted until *after* the move. The vessel is allowed to return fire with all guns that were able to fire when the move started.

(d) When firing at *wooden* ships, the same procedure as when firing at Ironclads is followed. However, shots that *do not* have Special Effect (those that strike the hull) cause points damage according to the weight of shot (Col 2-Chart A) and not according to Col 3 of Chart A.

(e) When firing at *point blank* range—the gunnery procedure is unchanged, but since there is no *miss* section on Chart B, complete misses are not possible.

CHART (B)

No of Guns	CARDS DRAWN ACE	2	3	S.B.M.! LONG RANGE MISSES (4 TO KG) 4	5	LONG RANGE MISSES → 6	7	8	MEDIUM RANGE MISSES → 9	10	JK	CLOSE RANGE MISSES → QU	KG
12 GUNS	1	2	3	4	5	6	7	8	9	10	11	12	1
11	1	2	3	4	5	6	7	8	9	10	11	1	2
10	1	2	3	4	5	6	7	8	9	10	1	2	3
9	1	2	3	4	5	6	7	8	9	1	2	3	4
8	1	2	3	4	5	6	7	8	1	2	3	4	5
7	1	1	2	2	3	3	4	4	5	5	6	6	7
6	1	1	2	2	3	3	4	4	5	5	6	6	1
5	1	1	1	2	2	2	3	3	3	4	4	5	5
4	1	1	1	1	2	2	2	3	3	3	4	4	4
3	1	1	1	1	X	X	2	2	2	2	3	3	3
2	1	1	1	1	1	X	X	X	X	2	2	2	2
1	X	X	1	1	1	X	X	X	X	1	1	1	1

(f) It will be observed that some guns are capable of firing either shell *or* solid shot. If it is intended to fire solid shot (which has certain advantages under some conditions) clear attention must be drawn to this fact—either by statement of written observation. If no such attention has been made, it is assumed the pieces concerned were loaded with explosive shell.

(g) Ironclads are unarmoured at the bow and stern, and shots able to strike this portion of the ship (i.e. if the guns can bear upon it) will have the same effect as if fired at *wooden* ships (see page 71).

(h) The angle of fire of the broadside guns is 30 degrees which is to be taken from the centre of the model. If it is disputed whether a vessel can be fired upon, it is quite easy to settle the point by use of an angled card (see illustration, page 37).

SEQUENCE OF EVENTS DURING A GUNNERY MOVE

The sequence of events also describe the use of the various charts. If a first reading gives the impression that the system is very complex, it is not so in practice. In fact, the method was devised after much experiment so that all the important factors can be accounted for in a system that is reasonably simple and fast to operate.

Most of the results of firing guns will not go beyond the stage of using charts A and B. Only a few special effect hits will require the use of other tables.

After a while, it will not be required to follow the sequence since a couple of practise battles will commit the whole thing to memory.

1. Each player takes it in turn to fire the guns of one warship at a time. This will only occur after all ships have been moved.
2. Measure range to target(s).
3. Having found the range, consult the *Gunnery Information Chart* (A) to determine whether this range is listed under

Long range, Medium, Close, or Point-blank.

Simply find the calibre and type of gun that is being fired in Column 1, and find the range in one of the columns 4-5-6. Above this square atop the chart is shown the range rating.

If the range is *less* than the minimum figure shown in the Close range section, then the range is taken as *point-blank*.

4. The rifled guns aboard ship are fired first followed by the Smoothbores. The Smoothbores are fired *twice*.
5. Cut one card for each set of rifled guns carried, i.e. if both 8 in. and 7 in. guns are carried, fire them separately.

   Hits are determined by the use of the *hit* or *miss* table.

(B) Assume that 15-7 in. rifled guns are firing a broadside. In this instance fire 12 guns followed by 3 guns.

Find number of guns firing on the extreme left of Chart B. Find card drawn on top line. Where the two lines intersect, locate number of hits scored.

Care must be taken to look for total misses. Study the labelled arrows above the chart which are self explanatory. The arrow indicates that when firing at Medium range, Cards 9, 10, JK, QU, and King are *Misses*.

6. Each time a set of guns are fired and hits are registered, follow by casting one dice per hit scored. This is a quick procedure because if 5 hits are scored, simply throw 5 dice together. On odd occasions when large numbers of hits are scored—cast two lots of dice, i.e. 5 dice followed by 3 dice if 8 hits are registered.
7. Only dice that show 1 are to be registered as *special* effect hits. All dice showing 2-3-4-5-6 are non penetrating hull hits and score minor damage according to Column 2 of the *Gunnery Information Chart* (A). Record the points damage scored in the damage section of the enemy ship hit by these shells.
8. Now deal with hits which registered 1 on the dice. Obviously these will be very few. In most cases all hits will be non-penetrating hull hits.

Cut one card per Special Effect hit-using Chart (C) and follow any further sequences from there. Since only a few hits will qualify for the use of this chart no attempt will be made to follow this sequence here.

However, any damage resulting from Special Effect hits (Fires, Explosions, Points damage etc.) must be entered as they occur.

SPECIAL EFFECT HITS

As indicated in points 7 and 8 of the Gunnery sequence, only hits which have been followed by a dice throw of 1 are Special effect hits and qualify for the use of these charts and tables.

Card	*Long range*	*Medium range*	*Close range*	*Point-blank*	
				Rifled	*S.B.M.*
Ace	Deck	Deck	Deck	Deck	Deck
2	Deck	Deck	Deck	Deck	Deck
3	Deck	Deck	Deck	Pen	Pen
4	Deck	Deck	Pen	Pen	Pen
5	Deck	Deck	Pen	Pen	Pen
6	Deck	Pen*	Pen	Pen	Pen
7	Deck	Pen*	Mast	Mast	Pen
8	Pen*	Mast	Mast	Mast	Pen
9	Mast	Mast	Mast	Mast	Mast
10	Mast	Rigging	Rigging	Rig	Mast
JK	Rigging	Rigging	Rigging	Rig	Rig
QU	Rigging	Rigging	Rigging	Rig	Rig
KG	Rigging	Rigging	Rigging	Rig	Rig

*S.B.M. Guns cannot penetrate armour beyond close range. Count as non-penetrating hull hits:

Mast Hit

Loss of 1 knot speed and loss of control for number of short moves denoted by dice roll.

When a ship loses control it is advisable to afford her maximum protection for she is then highly vulnerable to ramming.

Rigging Hit

Assumes a single sail shot away. Loss of control for one

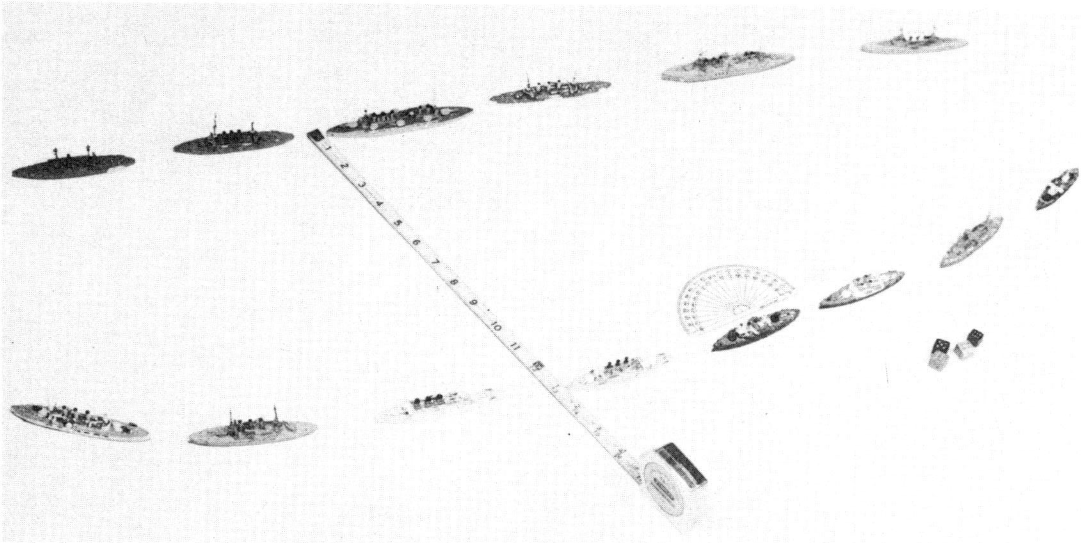

Line of Battle, *circa* 1909. Two lines of pre-dreadnoughts shoot it out. On this scale, one mile is equal to 60 inches and thus the lines are in reality, too close together. Models by Navis, Mercator, Hansa, plus 'Danton' class battleships built by Ron Hughes and Barrie Richardson.

short move if dice roll of 1, 2, is cast. Record number of sails shot away under damage control. Three sails destroyed equals loss of speed of 1 knot.

DECK HITS (cast single dice for damage)

	Shell	*Solid shot*
1	×2	×2
2	×5	×3
3	×10	×4
4	Fire (500)	×5
5	Fire (500)	×6
6	Fire (1,000)	×8

×2, etc., means times weight of shot (see column 2, Chart A).

Fire damage is shown and the fact that a fire is burning must be noted under damage control. Draw a miniature fire. Attempt to put out the fire each 1 minute move. Failure to do so will lead to the fire burning on with repeat of initial fire damage or to an explosion. If fire is put out, record this fact by crossing out fire drawing. *Explosion* 1,500 points damage per number on dice, 6 is 10,000.

FIRE FIGHTING

1, 2, 3	Fire out	
4, 5	Burns on	(original damage repeated)
6	Explosion	

See also optional rule concerning fire fighting by withdrawing from action.

PENETRATION EFFECT

Use 2 dice, adding scores together

Throw	*2*	*3*	*4*	*5*	*6*	*7*	*8*	*9*	*10*	*11*	*12*
Shell	E	E	×5	×10	×15	×20	Fire	Boiler	B	Mag.	Mag.
Solid	E	×2	×3	×5	×6	×8	×10	×12	×14	B	Mag.

E means engines. B means boiler. M means magazine.

Hit in the engines denotes use of sail power only. Hit on boiler means half engine speed only.

Magazine hit means complete destruction of vessel.

The Austrian attack at the Battle of Lissa, 1866 (ironclads shaded). There is nothing clever in this attack formation which appears to have enabled the Italians to cross the 'T' of the Austrians. Ironclads were almost immune to gunfire in this period however.

CHAPTER 6 Pre-Dreadnought Gunnery System

The system explained in this chapter relates to gunnery only and it is most suitable for the pre-dreadnought period when only a few heavy guns were carried.

It is a fairly complicated system that accounts for many of the important points related to gunnery. As such, it is really only suitable for very small actions. If you desire to fight actions using numbers of pre-dreadnoughts, then the World War Two rules are quite suitable. The main difference between the period 1880 to 1945 is simply increases in rates of fire, gun ranges, and so on. You need only modify the WW2 rules in places to make them suit the era of the pre-dreadnought.

This system is a means of pin-pointing the exact location of each hit. The sequence of getting a shell not only to hit but also to go through the ship's armour, is a fairly lengthy one and takes into consideration as the main factor, efficiency of the gun in question. The actual method of computing ship damage is a fairly simple point system apart from the vital hits. However, an elaborate method can easily be applied by those desiring it so that actual damage will result because of *where* the shot has fallen. Some ideas are mentioned later.

Models can be used to fight the actions, although a simple but quite satisfactory idea is to simply draw the ships on a large sheet of paper. Work out a scale for gun range (i.e. 1 in. equals 1,000 yds.), and simply redraw the ships in a fresh position according to the duration of the move.

Most battlegamers will probably prefer to use the little scale 1-1200 model ships and get a visual impression of the battle. Unless a much smaller scale is in fact used, a properly scaled gun action is not possible. To fire 12,000 yards we would need measure off 480 ins. which is the sort of distance that requires the use of a large hall.

In a single ship engagement, the fact that the range is is reduced from 480 in. to 60 in. is of little importance. But when two lines of model ships are in action, then the reduction in field of fire poses problems. In effect the models should be one eighth of actual size. Still, this need not spoil our game. I suggest that range *and* speed is reduced to one eighth of actual scaled distance. Thus you reduce a 16-knot ship to 2 knots (two inches per move). The problem can be overcome in part by allowing the models to steam half an inch apart instead of two or three lengths which is the proper station. When torpedoes are fired by a method representing miniature torpedoes, we actually move only one in eight in order to get the hit ratio correct. But the models are still too far apart and arcs of fire and gun ranges are not accurate as a result so various other compensations will need be made by the purists. Others, who appreciate that a good game can still be had despite the faults, will not be so particular.

METHOD

This method involves drawing a simple diagram of the warships involved in the action. Most readers I assume, will either have or know of the recent re-prints of well known naval reference books. Weyers and Janes have both been reprinted for the years 1905, 1914, and 1919. It is the 1905 edition that will interest us. The reprints are relatively inexpensive. These are but two sources. Brassey's naval annual is another and there are various other publications that have one thing in common.

All of them display an outline drawing (besides photos) of each ship portrayed therein. Shaded into the outline are the areas protected by armour and the exact thickness of this armour at various points is clearly indicated. It is that which is so useful to us.

The simple plan is copied and enlarged. We only copy

the basic drawing and do not need to show armour distribution. When we need to know the thickness at a given point, we simply refer to the drawing in our reference book.

The copying of the plan requires no special skill. Rather, it is more important to work carefully. The plans themselves are only basic drawings of the warships and do not show detail. They are not supposed to be highly accurate either but they do give us the main lay-out and the proportions of the vessel will be quite accurate enough for our purposes.

The diagram you draw does not need to be any particular scale. The easiest way is to simply measure the length and depth of the drawing in the book and then double or treble the measurements when you make your copy. Judgement of eye will be necessary when drawing the superstructure and turrets etc. Some parts can be measured though and then enlarged twice or three times as was done with the hull.

The diagram should cover two lines of graph blocks on a sheet of graph paper. Any good stationer can supply graph paper in book form and this is ideal for our purpose.

The superstructure of your drawing should encompass one line of graph blocks and the hull should be drawn upon that below. The illustration will show this. It has not been possible to show the individual squares in each block (there are 100) but the outlines of the graph blocks are indicated.

Whatever warship is drawn, your diagram must show magazines, engine room and boilers and this should be done as in the illustration. The main armoured belt never covers this area. That is because it is below the waterline and shells falling at an angle must pass through the belt in order to plunge into these vital spots. For the purpose of our game we 'extend' the main belt to cover this area because we are simply locating bursting point and not accounting for angle of decent of projectile.

Here is how we fix the exact point at which each shell has hit the target. The graph blocks above our drawing are marked by card numbers or court cards. If the drawing is 9 blocks long, number them Ace to 9 and leave out 10 to

King (draw again if one of these cards are cut).

If the drawing is 5 to 7 blocks long, each block can be marked with two cards (see illustration). These cards indicate which column of the drawing is hit. The two *lines* of blocks (ship's length) are simply marked Red and Black. Red means the shot must fall upon the superstructure turrets, or deck. Black means the shell must hit the hull.

Thus if we draw a 9 followed by a red card, this shows our shot will fall within the upper graph block around *Patries* superstructure and forward funnel. This is where the cards intersect.

Now we need to determine which square among the one hundred within the block is the one where our shell has hit. We draw a further two cards. If the first card is black, we double the number shown on the table below against the second card drawn. If it is red, we count the number shown and do not double it.

	Hearts	*Diamonds*	*Clubs*	*Spades*
Ace	1	14	27	40
2	2	15	28	41
3	3	16	29	42
4	4	17	30	43
5	5	18	31	44
6	6	19	32	45
7	7	20	33	46
8	8	21	34	47
9	9	22	35	48
10	10	23	36	49
J	11	24	37	50
Q	12	25	38	—
K	13	26	39	—

An easy way to find any square within a block is to count 10, 20, 30, etc. downwards on the last line of squares in the block. Thus square 93 is 9 lines down plus three squares from the opposite side (i.e. first line).

It will be helpful to copy this chart so that it can be permanently displayed on the table without having to find the page in this book. This applies to other charts and tables.

What if a shell should strike a spot on the red line that is blank? In this event, simply bring the hit 'straight down' (on the same line) until part of the ship is reached. The first complete square to be reached is deemed to be the point of explosion (as first hit shown on the illustration).

Once we have located the exact spot on our diagram, we refer to the diagram in our reference source and we can see whether the shell hit at a point protected by armour. We can also see the exact thickness of any armour at that point and then we need determine whether or not our projectile has burst through it.

First we need to establish the quality of the guns of each warship. Since this is a factor that also has a direct bearing upon the odds of hitting, the quality of the gun in question will be known prior to any shell hits.

All the reference books previously mentioned also include gunnery and torpedo tables. By reference to details contained on the same page as the drawing you can find the model or mark of the gun and locate it on the gunnery table.

The most relevant information is rate of fire of that gun and its penetrating power. Penetration figures are usually given at 5,000 yds., 3,000 yds. or both. Under the diagram you have drawn, mark down the penetration figure of the guns at one of the figures given (i.e. 14 ins. at 5,000 yds.).

By using the armour penetration chart (see page 89) it is now possible to determine the power of penetration at *any* range. Suppose your gun can pierce 9 ins. of armour at 5,000 yds. and this is the only figure you have. All that need be done is to find the column where the figure 9 ins. intersects the 5,000 yd. line. That column then shows penetration figures at all other ranges (6 ins. at 7,000 yds., 15 ins. at 1,000 yds. etc.). The table is not complete. If you are firing at a figure like 5,750 yds. you would need calculate penetration between the figures given for 5,000 and 6,000 yds. Likewise, the figure given in the reference book may be 14½ ins. penetration at 5,000 yds. To find the armour that can be pierced at (say) 9,000 yds., you would need take the figure half way between that listed for 14 ins. and 15 ins. at

5,000 yds. The answer at 9,000 yds. in this instance is 8½ ins. (see chart).

The penetration chart also serves as a simple but effective gunnery table (for scoring hits) and hitting odds are based upon the quality of the gun. The old 13·4-in. guns of H.M.S. *Royal Sovereign* will penetrate less armour than the 12-in. guns of the more modern French battleship *Patrie* and hence will also be less likely to hit the target. This is true to form.

When you 'compute' the penetration figure of your gun according to its capabilities and the range at which it is firing (as already directed) you arrive at a figure given in inches. The number will be anything from 1 in. up to 26 ins. As regards odds of hitting the target, the penetration figures are converted into a card table which starts from ace (1 in.) up to black king (26 ins.).

HITTING TABLE

Pen figure	*Red card*	*Pen figure*	*Black card*
1 in.	Ace	14 ins.	Ace
2 ins.	2	15 ins.	2
3 ins.	3	16 ins.	3
4 ins.	4	17 ins.	4
5 ins.	5	18 ins.	5
6 ins.	6	19 ins.	6
7 ins.	7	20 ins.	7
8 ins.	8	21 ins.	8
9 ins.	9	22 ins.	9
10 ins.	10	23 ins.	10
11 ins.	JK	24 ins.	JK
12 ins.	QU	25 ins.	QU
13 ins.	KG	26 ins.	KG

If the penetration figure was only 1 in. you would need draw a red ace in order to score a hit. One card is drawn per shot fired. If the figure was 3 ins. then any card up to Red 3 is a hit. All others are misses. Likewise a penetration figure of 9 ins. would mean that any red card from Ace to 9 is a hit. If the figure exceeds 13 ins., you move onto the black card column. If your gun can penetrate 21 ins. at the range to the target, *all* Red cards *plus* black cards from Ace to 8 are a hit. Should the figure transpire to be part or fraction of an inch, count the card above if the part is one half or higher, i.e. $7\frac{1}{2}$ ins. a card up to 8 may be drawn for a hit.

LONG RANGE FIRE

When ships are firing at long range, it is hardly possible to hit the side of the ship. Shells are much more likely to plunge onto the target and the following rule is in force.

When ships are firing at the last quarter of their maximum range, all shots that hit are deemed to be plunging shells. Locate point of shell hit as instructed, but it is no longer necessary to go through the side armour.

If however the shell strikes the vital areas, it must be capable of piercing the ship's deck armour which is located immediately above the vital area. If it fails to penetrate, it is deemed to have exploded above it (on the same line).

Whatever system is used, penetrating shots should on average cause far more damage than those that do not. On some occasions however, non penetrating hits can cause bad damage. At the Dogger Bank battle in WW1 it was a German shell that failed to pierce H.M.S. *Lion's* armour that knocked her out of the fight (it displaced the armour).

Now it should not be an absolute rule that a shot either will or will not penetrate according to any penetration table. At Jutland (1916) some shells 'broke the rules' as it were and got through German armour at ranges where they were not supposed too. And others failed to go through when they should have done so.

The following two rules are optional and need not be included. They will increase the sequences needed to go through a gunnery move. They would also add a touch of realism.

1. The first point is whether the shell manages to exceed (or otherwise) its nominal penetration figure at ranges given. Throw two dice. One Red and one Yellow.

Red Dice

1 to 4	As table	
5	Increase	Consult Yellow Dice.
6	Decrease	

Yellow Dice

1,2	one-fifth
3,4	one-quarter
5,6	one-third

Example: If normal penetration is 15 ins. and the decrease figure is one quarter then the relevant figure for that shot is now $11\frac{1}{4}$ ins.

2. This is a point for which I shall provide no rules. It is not vitally important and it would be easy enough for individuals to formulate rules if they wish to do so.

It concerns the angle at which a shell strikes the side armour. This is easily seen by simply noting the angle the shell must arrive at according to position of ship firing and target. I am not talking of angle of descent (which depends on range) but of that illustrated below. Obviously, as the angle between A and B becomes more acute, penetration odds decrease.

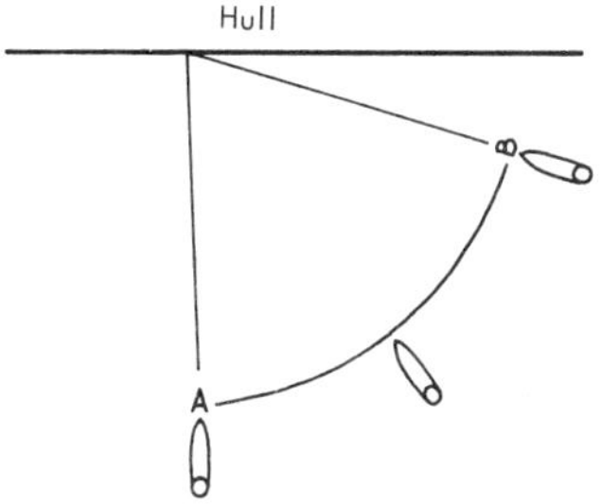

A further factor which some may wish to account for is the fact that smaller targets are more difficult to hit. This can be put into practice by having saving throws for ships smaller than a certain size. Saving throws are made by the players whose ship is 'hit' and follows when his opponent draws hit cards. A throw of 1 may save a shot hitting an armoured cruiser and 1 to 4 of hitting a T.B.D.

DAMAGE

This can be made simple or complicated. A hit on the funnel might merely cause points damage or it may have results *because* the funnel is destroyed (smoke nuisance etc.).

The ship can be sub-divided into compartments and a value placed on each compartment according to the tonnage of the vessel. Hits can cause varying damage. A lucky hit by a large shell might destroy several compartments. When a third of the compartments are destroyed, we cut cards to see if the vessel will sink and thereafter when additional ones are destroyed if she does not.

Hits in the magazine, engine room etc., should not mean automatic destruction of that part. A 9·2 in. shell that bursts in an engine room is less likely to render the ship stationary than a much larger shell. A reduction in speed according to the size of the shot seems the best solution.

A simple points system is as follows (but hits on turrets and vital areas can still cause relevant damage. A shot must penetrate the turret's armour in order to destroy it).

For each 1,000 tons award 200 points.

Shells score as follows:

13·5 in.	250
12 in.	170
11 in.	120
10 in.	100
9·2 in.	76
8 in.	50
7·5 in.	40
7 in.	32
6 in.	20
5 in.	12
4 in.	6

Weights of projectiles are always given in the reference books mentioned earlier. To apply other calibre guns to this system, strike off last figure of shell weight and double remainder, i.e. 664 lbs.=66=132 p.v.

The number of shots required to destroy the target is approximately correct.

Speed should fall when over two-thirds of the points are lost unless already dictated by engine room damage.

Shell hits can cause standard damage or if the sequence is carried one step further dice can be thrown to see if the explosion causes above average damage or less. Here is one method.

Dice throw	*Pen hits*	*Non Pen hits*
1,2	Normal	$\frac{1}{4}$ damage
3,4	Double	$\frac{1}{8}$ damage
5,6	Half	nil

This is a simple method that can be made more sophisticated if desired.

RATES OF FIRE

Assuming that we use the oft quoted 1-1200 model which, moving at a scaled speed of 1 in. per knot per minute, we would find the ships firing every few inches if actual rates of fire were used. It is necessary to 'slow down' the rate of fire and this is in keeping with the gunnery table which would otherwise make hitting too easy.

Rates of fire is 6 times slower than normal. *Normal* rate of fire is indicated on the left and opposite is shown the rate of fire relevant to this system.

Rounds per minute	
·5	1 shot in 12 minutes
1	1 shot in 6 minutes
2	1 shot in 3 minutes
3	1 shot in 2 minutes
4	1 shot in $1\frac{1}{2}$ minutes
5	1 shot in $1\frac{1}{4}$ minutes
6	1 shot in 1 minute

The battleship *Patrie* has a rate of fire of $1\frac{1}{2}$ rounds per minute and hence fires every $4\frac{1}{2}$ minutes in our game.

All rates of fire below are 1 shot per minute plus one added shot per clear number of guns firing indicated.

7	Add 1 shot in 6
8	Add 1 shot in 5
9	Add 1 shot in 4
10	Add 1 shot in 3
11	Add 1 shot in 2
12 r.p.m.	×2

If R.O.F. is 7 r.p.m. and 5 guns are firing, no shots are added. If however, 6 guns fire, a seventh shell is fired (draw for 7 hits).

If the R.O.F. is 10 r.p.m. and 6 guns are firing, add 2 shots and draw cards for up to 8 possible hits.

It is difficult in practice to remember when it is each ship's turn to fire. The time scale shown here is used to note when each ship has fired so that it can be seen when her guns are reloaded. Moves are usually in terms of one minute but guns firing between minutes will still be reloaded quicker than those firing on the minute, i.e. ship firing

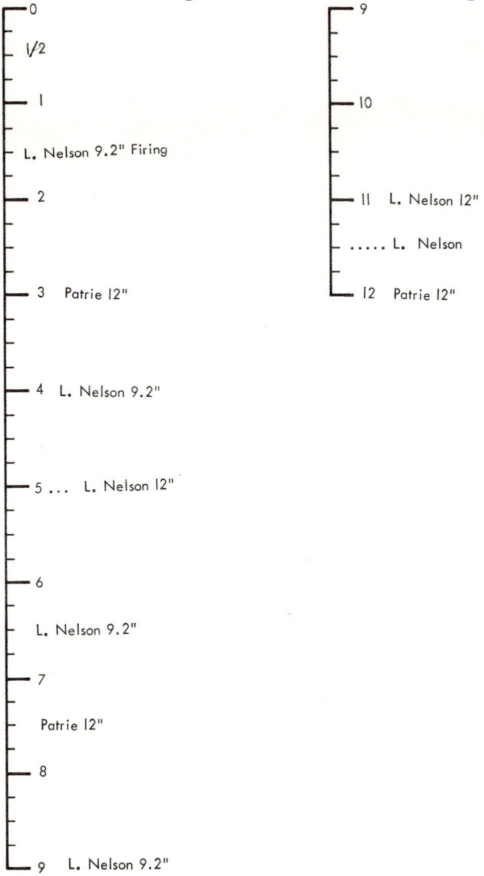

every 1½ minutes compared to another firing every 2 minutes. The example is taken from a battle between *Patrie* and *Lord Nelson*. The *Patrie's* 6·4-in. guns were out of range.

ARMOUR PENETRATION CHART

Range	*Penetration figures in inches*											
14,000 yds.	½											
13,000 yds.	2	1	½									
12,000 yds.	4	3	2	1	½	½						
11,000 yds.	6	5	4	3	2	2	2	1½	1	½		
10,000 yds.	8	7	6	5	4	4	4	3½	2½	1½	½	
9,000 yds.	10	9	8	7	6	6	6	5	4	3	2	1
8,000 yds.	12	11	10	9	8	7¾	7½	6½	5½	4½	3½	2½
7,000 yds.	14	13	12	11	10	9½	9	8	7	6	5	4
6,000 yds.	16	15	14	13	12	11¼	10½	9½	8½	7½	6½	5½
5,000 yds.	18	17	16	15	14	13	12	11	10	9	8	7
4,000 yds.	20	19	18	17	16	14¾	13½	12½	11½	10½	9½	8½
3,000 yds.	22	21	20	19	18	16½	15	14	13	12	11	10
2,000 yds.	24	23	22	21	20	18¼	16½	15½	14½	13½	12½	11½
1,000 yds.	26	25	24	23	22	20	18	17	16	15	14	13
	Heavy guns, 11-in. upwards							*10in.*		*9·2in.*		*8in.*

Part Two

8,000 yds.	1½	½			
7,000 yds.	3	2	1½	1	
6,000 yds.	4½	3½	2½	2	
5,000 yds.	6	5	4	3	1
4,000 yds.	7½	6½	4¾	3¾	1¾
3,000 yds.	9	8	5½	4½	2½
2,000 yds.	10½	9½	7¼	5¼	3¼
1,000 yds.	12	11	8	6	4
	7·5in.	*7in.*	*6in.*	*5in.*	*4in.*

The gun calibres listed below each chart, merely indicate that the figures listed above them correspond very approximately to a gun of their power. When penetration figures are not available, they can be taken as such. It is important to note, however, that the actual figures can vary a good deal from one model or mark of gun to another (of same calibre).

GUNNERY SEQUENCE

1. Check range to target.
2. Check to determine whether the shots are falling at the last quarter of their maximum range. If so, shells do not need to pierce side armour (see rules).
3. Find card necessary to hit target. This is according to the range and penetration quality of the gun(s) being fired.
4. Draw one card for each shot being fired. Mark down number of shots registered as hits.

HITS

5. Cut two cards for each hit to indicate which graph block of 100 squares on the diagram has been hit. Then draw two further cards to pinpoint the exact square within that block where the shell has deemed to explode.
6. Before dicing to see the damage caused, locate exact point on the diagram in your reference book to see the thickness of armour at the point struck by the shot. Calculate whether the shell can penetrate the armour (if any) at that spot. If using the optional rules, calculate angle shell hits ship and any increase/decrease in normal penetration.
7. Having determined whether the shell has or otherwise penetrated side or deck armour, throw dice to calculate the damage caused.
8. Mark or write down this damage and if using a points system, add to any previous damage.

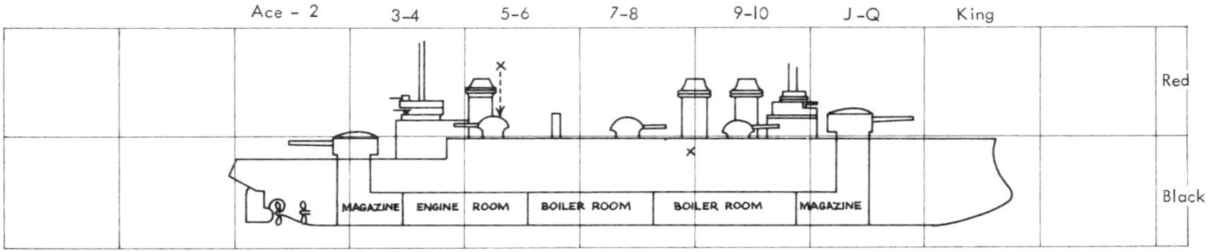

FRENCH BATTLESHIP "PATRIE" OF 1902

Hit (1) on roof of 6l5inch gun turret must penetrate 6" of armour in order to destroy it.

Hit (2) has hit at point unprotected by armour and must enter ship.

1st. Hit

Cards Drawn

(1) Red

(2) Red

(3) 7 of Clubs (sq. 33)

2nd. Hit.

(10) Black

(2) Red

(3) 7 of Hearts (sq. 20)

CHAPTER 7 **Map Campaigns**

When a naval wargamer has become experienced at fighting set-piece battles, he invariably becomes interested in waging a campaign and in this field, since the potential is absolutely unlimited, his interest need never become jaded.

I shall describe three types of map campaigns. These are as follows.

1. *The self contained game*
   This is the easiest sort of map campaign to set up. It does sacrifice some realism but it enables two players to wage large scale warfare. Models are not used at all and both campaign and battle moves occur on the same map. The hypothetical 'world war' game described in a later chapter uses this principle.
2. *The semi-secret campaign*
   This method uses exactly the same principle as the self contained game. The only difference is that the symbols (flags) need not represent an individual vessel but a whole group of them. It is the nature of the force represented by the flag that is kept secret. The battle side of the game can be transferred to a separate map or else models may be used.
3. *Completely secret game*
   A third person is needed to act as umpire and to watch each player's maps for contacts. It is rather more involved but also very realistic and the maps are only used for campaign purposes. Battles are fought by using models. Symbols are not used since the progress of any force is represented by drawn track lines.

Maps need not be used only to fight whole campaigns. In fact it is best to start with something simple. A typical problem might be that of escorting a convoy to a specified area in the face of heavy enemy opposition.

When moving sailing ships upon a campaign map it is best to dispense with elaborate sailing rules and allow vessels to move at varying speeds according to wind direction and strength, i.e. instead of attempting to tack against the wind, allow a very low speed in the desired direction. More elaborate rules ought to be confined to the models.

PREPARING THE MAPS

All three games make use of large graphic sheets upon which the forces are moved. Whenever squares are used to represent distance an immediate fault is apparent. Anything moving diagonally across the map will cover a greater distance than another moving in a straight line and covering the same number of squares.

Provided this fault is realised and accounted for it need not prove a serious matter. To offset this inaccuracy, the graphic sheets eliminate the need for measuring and great distances can be calculated at a glance because graph sheets are sub-divided into blocks tens squares across.

The completely secret game (3) does not use the squares except as reference points and track lines must be measured.

These large graphic sheets can be obtained at your local art shop or stationer. The sheets are divided into one inch graphic blocks each containing 100 squares. There are 29 blocks across and 21 downwards so that each sheet consists of 609 graph blocks and 60,900 individual squares.

One sheet can thus represent a fairly large area but 4, 6, or 8 can be joined together to make truly huge areas.

When you purchase the map also ask for cartridge paper the same size to which will be stapled the graphic sheet. Then with sticky tape we can secure a loop of string to the backing. The completed map can be hung in the same fashion as a mirror, or it can be stuck to the wall with sticky tape around its edges.

It will be necessary to locate individual squares from time

to time and so we must number the blocks across the map, and letter those downwards.

This enables us to make the following simple code that will locate any square in a few seconds. 27/E Square 75.

The numbers and letters are boldly marked around the edges of the map. Individual squares are easy to find as there are 100 in each block. The first square is No. 1 and the last is No. 100.

THE SELF-CONTAINED WARGAME

Everything happens on the operational map. We do not use models or transfer to a separate battle map when contact is made between rival warships.

There is no secrecy because the disposition of the enemy forces can be seen at a glance. The advantage of this game is that it is easy to prepare and quick in operation. We do not attempt to recreate accurate battle tactics or time and distance scales. In fact, the game is simplified all round so that we can concentrate on waging warfare on a large scale. Air power and land forces can be included provided that rules apertaining to their use are kept really simple.

A flag represents individual ships. These are made by simply sticking a pin through a small square of coloured card. So that various types of shipping can be easily distinguished, a separate colour is used for each type.

1st Rate	Battleship	Blue
2nd Rate	Heavy Cruiser	Red
3rd Rate	Light Cruiser	Green

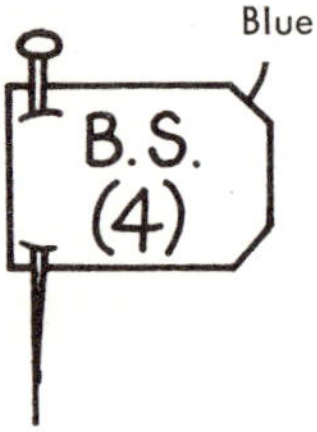

If we have four battleships etc. then we make four blue flags and mark them B.S.1, B.S.2, B.S.3 and B.S.4.

Our fleet list will show us that B.S.1. represents *Bismarck* and that B.S.2. represents *Scharnhorst*. That same record will indicate speed, gunpower, and all the other essential data concerning the ship in question. Speed can be marked on the reverse of the flag but this will soon be committed to memory.

The appearance of the map when operations are well under way is most impressive. It looks like those military maps to be found in the planning room of any H.Q. On these, little coloured flags are positioned to denote the whereabouts of forces engaged, in reserve and so on. Although secrecy is lacking in this type of campaign, it does have its compensations (and very few wargames *do* use secrecy) in that a glance at the map gives a sweeping impression of the whole activities going on in every part of the 'world'.

This enables us to coordinate operations and plan our future needs with just two operatives.

As I have made clear, the method is designed to merely give us the end result of a naval battle and to allow us to concentrate on grand strategy and manipulate large forces.

First, a single move will represent the passing of a day and each ship will cover the same number of squares as it has speeds in knots. When in action each ship fires one salvo but this salvo represents the damage a ship could do over a fairly long period of time (see battle rules).

It is possible that a battle may last for several moves but we must use our imaginations to assume that it was *not* a continuous affair that lasted for a few days! If you record a history of your war then it is practical to eliminate actual faults by writing the battle account as if it really happened. For example if a destroyer action lasted for 3 moves, record the battle as taking place on the first day and lasting for (say) three hours.

BATTLE RULES

The simple rules given here are intended for periods when the contestants fought a fair distance apart and did not melee or board. Since complicated rules are not a desirable factor as far as this type of campaign goes, they can be adapted for use with sailing men-of-war.

1. When both sides are in a position whereby either can move a ship into gun range of his opponent, they take it in turns to move single ships. Those unable to enter battle are left until last when they are all moved together.
2. The first ship to move can be fired at by every ship that has the range to hit her and at any point during its move and not merely at the end of it.

The first ship to move may do so perhaps 20 squares directly towards the enemy fleet. Now a visual impression of the map at this moment is that one vessel has departed suddenly from its companions and is determined to commit suicide!

In fact, we must remember that all the ships are really moving together and we must wait until the end of the move to get the true picture of what has happened.

Of course, the first ship to move could suffer the concentrated fire of a mass of enemy ships and be blown asunder in short order. In practice this would not be very wise since they would not then be able to fire at ships following her (i.e. moving afterwards). The attacker does however, suffer a disadvantage. Once a vessel has moved it has committed itself to concentrated fire if the enemy decide to destroy her. On the other hand, the other side can withdraw ships as they are damaged and hence need only be brought under fire by one enemy ship before retiring.

Such an assault would usually only occur upon an inferior force thus nullifying many of the points mentioned above. It is more usual for both sides to close the range slowly. If one side is in full flight then the situation will not arise at all.

Despite its obvious faults it is surprising how this simple method of ship fighting gives us a very fair degree of realism. It is admitted that the manner in which the ships (flags) manoeuvre on the map is a little odd compared to a force of real ships, yet the end result of any map battle is always authentic. This is what we are aiming at. After all, if we have a fleet of a hundred ships or more, we really must compromise somewhere.

If it is intended to fight a 'modern' campaign, then there are battle rules elsewhere in this book that although written mainly for model use, are easily adaptable for use with the self contained game.

The exception is when surface torpedo attack occurs for then we must use chance cards instead of the system of skill used when torpedoes are represented by markers (as used with models).

Remember that surface torpedo firing only takes place when gunfire is over so that only destroyers surviving any gunfire may fire torpedoes.

Ranges in use for guns and torpedoes need only be scaled in relation to each other. Apart from that, ranges selected should be a matter of convenience and suitability rather than any attempt to scale things accurately.

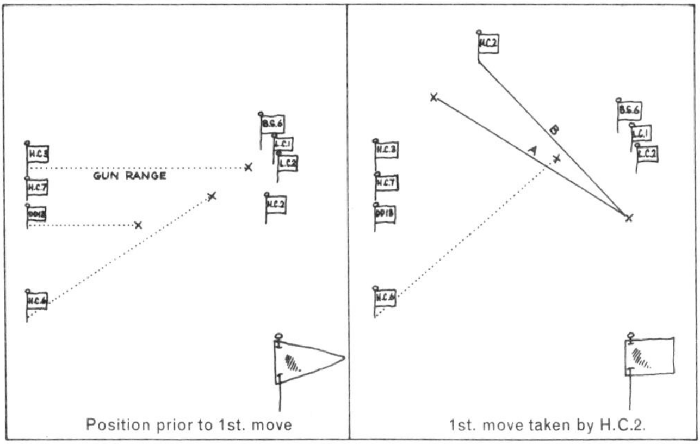

Heavy cruiser No. 2 has taken the first move. She steams into gun range of enemy heavy cruisers 7 and 13, but outside that of No. 6. Had she steered a course represented by the track-line A, she would have come under fire from heavy cruiser 6 also.

THE SEMI SECRET CAMPAIGN

The principle of this game is the same as that previously described.

The main essential difference is that a single flag can represent any number of ships from one of the smallest type right up to an entire task force. In this respect, less flag making will be involved than when preparing the self contained game. More complication is involved when rival forces make contact for then we must fight any actions away from the campaign map.

Whatever ships are used in your campaign, the first procedure is to make a fleet list. A thick exercise book is ideal. Enter the most important ships first (Carriers and Battleships) and do so in numerical order. Aircraft Carrier No. (1) will be on the first page followed by Carrier No. (2) and then number (3) and so on. Then when all the Carriers have been entered, you start on the Battleships followed by the Heavy Cruisers and so on. Each type of warship is numbered from (1) right up to the total number of that type that are included in your fleet. Therefore you will have a Carrier No. (1) Battleship No. (1), Heavy Cruiser No. (1) and so on.

If you are fighting another period then the same principle applies to whatever kind of ships are involved. When Napoleonic warships are being used, enter the 110 gunners first followed by the 100 gunners etc.

Each flag that is made bears a code letter and reference to your records will show exactly what that code letter represents. All this is worked out before the game starts. A flag might be a single frigate or a force of 100 ships. This then is semi-secrecy. The enemy can see the position of your forces and where they are heading but he does not know what they consist of until contact is made or they are sighted by submarine, aircraft, etc. The game can be improved by using a portion of dummy flags. This means we might see Nelson chasing a phantom French fleet which of course he did in real life due to receiving false information.

The dummy flags will add confusion and uncertainty—

surely two of the most common ingredients of naval warfare.

The game can include a proper scale and order of movement to match. I will leave this for the player to calculate. It will depend upon his individual requirements and it is not difficult to work out.

A relationship between the map scale and the battle area should be worked out. For example, Map scale 3 squares equals 8 miles. Floor scale one mile equals 10 inches.

Now presume current visibility on map is 16 miles (6 squares). Two forces sight each other. Models will be laid down 160 ins. apart.

These figures are merely examples.

Simple visibility rules are in order for this game. I suggest one card is drawn at the beginning of each day. The visibility is in miles according to the number on the card, (Jack is 11 miles). When Spades are drawn, the figure is doubled.

If the visibility is bad and force X wishes to engage force E (assuming both are not known) then X would need steam into visibility range which we shall assume is a mere 3 miles. This is somewhat risky since E might be more powerful. The models would be placed on the floor at a distance whereby a point blank action would ensue as far as modern warfare goes.

Here we can draw a parallel situation by reference to the battle of Jutland 1916, when Admiral Scheer, while chasing a British force in misty conditions suddenly found himself in a vast trap with overwhelming numbers of dreadnoughts (nearly forty!) firing at him.

Complete surprise is not possible because we can at all times observe the coded flags, but an element of being taken unawares and miscalculating the enemy's strength is always possible.

Identification from the air and by submarine is seldom accurate. Cruisers are often mistaken for battleships and so on. When aircraft or submarines sight surface units, it is only necessary to admit the total number of ships present and those that are large or small, (3,000 tons or less is classed as small).

Now let us suppose that a force is sighted and that its composition and general characteristics are revealed. In real life, having lost sight of the force, we do not know exactly where it is or whether it has remained as a compact force. As far as the game is concerned, we only need follow the progress of the coded flag in order to keep observation upon the identified force.

However, there is a simple solution to this problem. The force commander need only detach one or two individual ships or small groups of ships from the main force. These detached units will have separate code flags issued on their behalf and since the main force could have been recoded, then the enemy will not know which flag now represents the main force. The original flag may now represent one of the detached units.

At any stage of the game, ships may be detached from a group. All that is required is that a new coded flag or flags be made for the new group or individual ships created.

It is obvious that records of changes must be carefully and accurately maintained.

BATTLE

What happens when contact is made? There are two possibilities.

1. Models are used to fight the action.
2. Action transfers to a special battle map.

Little need be said here about the use of the models. The rival forces are laid out at a distance corresponding to that at which they sighted each other upon the map. If it is desired to get bearings and distances exactly worked out, then information on how to do this is contained in the next section and is to be found on page 110 and 111.

The bearing can be worked out in this instance by knowing the last two squares that the coded flag covered.

It is not essential in this type of campaign to have cruising plans of the ships represented by each coded flag. When contact has been made the models can be laid out according to the commander's wishes. It is perhaps not as satisfactory or realistic but it will save having to make these cruising plans.

The second method (2) makes use of a special battle map. This is simply a blank sheet of the graph paper used only for fighting surface actions. Those ships that have made contact upon the operational map are transferred to the battle map in the form of individual ships instead of code flags. The scale on the battle map is enlarged. If one square on the operational map represented 4 miles, then it might equal 1 mile on the battle map.

The battle is fought on the map according to the simple fighting rules given for the self contained map campaign.

Only rarely will battles occur off a coastline but when this happens, an enlarged coastline should be roughly drawn upon the battle map.

Let us now suppose that battle occurs and both sides lose a number of ships and then break off the action. By the time the fighting is over the contestants may have moved (say) 30 miles from the original positions. We now revert to moving the coded flags again unless certain ships have separated vast distances. A few calculations will show where the coded flag has to be re-positioned on the operational map. If a group of ships separated a great distance from the main force, then a new coded flag or flags would need be made for them. If we say that all vessels over 2 miles from the flagship must now be treated separately—then this will suffice. It is not difficult to relate battle map positions to the operation map when the scale is known of each, plus the original positions.

Drawings overleaf (pages 102–3) explain in picture form the principle of how this works.

Whenever ships are destroyed in battle, or when others leave a force to which they formerly belonged, then the records of which ships are grouped under the coded flags concerned, must be altered accordingly.

We must now consider another possibility. Suppose forces were rushing to join in the action and were not far distant from the original coded flags. Under such circumstances we

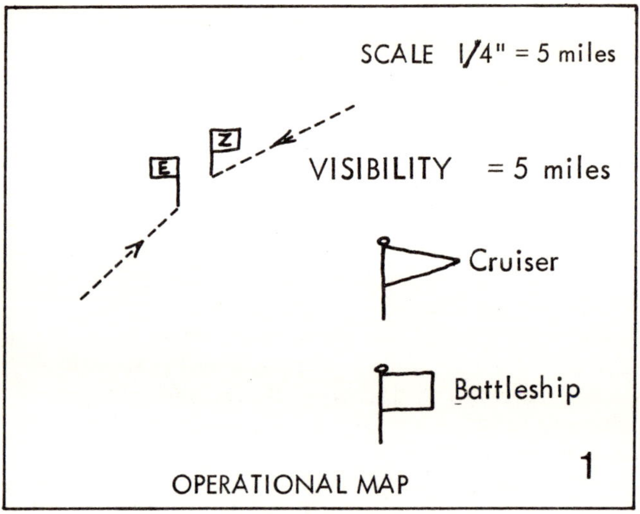

Force Z = Two Battleships

Force E = Three Cruisers

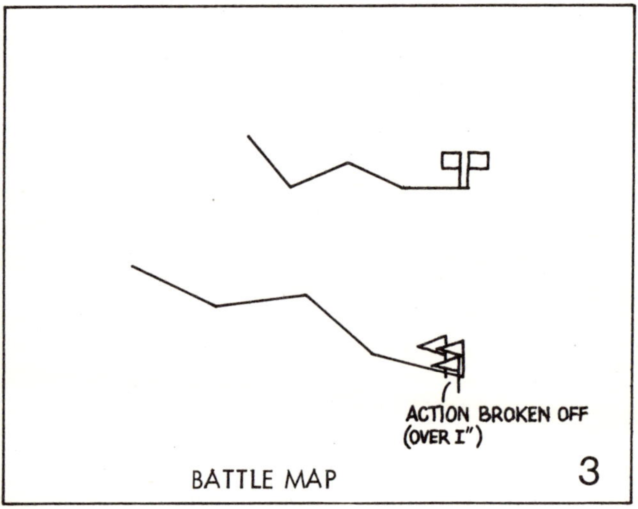

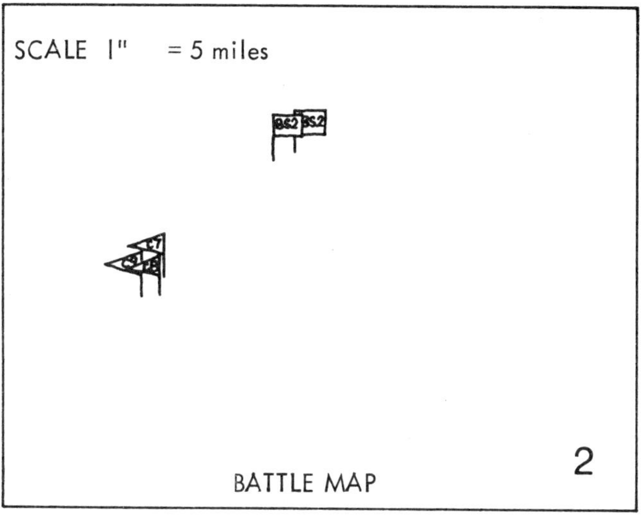
SCALE 1" = 5 miles
BS2
BS2
C7
C9
C8
BATTLE MAP
2

(Y)

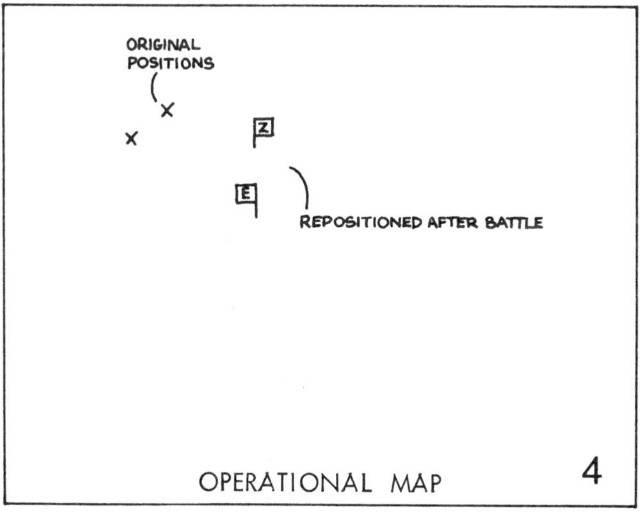
ORIGINAL POSITIONS
Z
E
REPOSITIONED AFTER BATTLE
OPERATIONAL MAP
4

would have to have a battle map move and an operational map move immediately afterwards.

We would need to work out at which point those ships on the operational map were able to join in the action and then transfer them (as individual ships) to their correct position on the battle map. All this will involve some fairly difficult calculations but it will come a lot easier with experience. What complicates the issue is the fact that the relative positions of ships in action to those out of visual contact (but nearby) on the operational map, is changing every move. The speed scale is different too and there is a possibility that an operational map move is normally a much greater period than a battle move duration. Despite all this, the problems can be worked out in a satisfactory manner by a process of calculation and logic.

The same thing applies when models are moved. Two forces may make contact on the map and a third may be close at hand and able to intervene shortly. Then we need to work out at what position on the floor the fresh force will appear and at what point in time. We have to remember that the ships have moved from their original positions and that where they are on the map will have to be calculated according to how they have moved from the original floor position.

If damaged ships rejoin a coded flag after the battle, then do not forget that a force must steam at the speed of the slowest ship therein. If it is not desirable to slow up the group of ships, then recode the damaged units and steam them individually.

Identifying enemy forces

There are 3 ways in which surface units can be located as regards modern warfare.

1. By aircraft. Ships can be seen at a great distance in good visibility but exactly what is seen can sometimes be misinterpreted.
2. By Submarine. The range of vision of a submarine is rather limited. However, a submarine has the advantage of locating the enemy without him being aware that his

position and presence have been revealed (unless it attacks or is located).

3. By surface vessel. The following table is an attempt to reproduce that which can be seen at varying distances.

20 miles	Smoke sighted.
15 miles	Able to determine whether smoke is from a few ships or large force.
12 miles	Number of ships can be counted.
10 miles	Number of large ships can be identified.
8 miles	Battleships can distinguished from cruisers etc.

A force that is moved to within 20 miles of an enemy ship or ships would be told that smoke had been sighted by the umpire when complete secrecy was being used. He could then make up his mind to investigate further by steaming closer. In bad visibility he may well come under fire directly the enemy is sighted. Regarding semi-secrecy, the table can only be used in part for obvious reasons. 'Smoke Sighted' does not apply.

COMPLETELY SECRET MAP MOVING

Tension, uncertainty, surprise, confusion and misleading interpretation of events happening elsewhere—these are the essential ingredients of naval warfare. At the battle of Leyte Gulf in 1944, against a fleet overwhelmingly more powerful and equipped with sophisticated radar and air search, the Japanese still managed to achieve surprise. By a simple ruse they caused the main American fleet to be in the wrong place so that the vital invasion beaches were wide open to surface attack.

This sort of thing cannot happen in a wargame unless we reproduce the element of secrecy. If we fail to detect the enemy, then he may strike a hard blow at us. We may detect him, but misconstrue his intentions with equally disasterous results. The destruction of the ill-fated convoy P.Q. 17 happened because the wrong conclusion as to the enemy's intentions were reached.

A wargame that reproduces these elements is vastly more exciting and nerve-racking than visual wargames. The rules and method described here is intended to make all the factors possible. I developed the system for games using upwards of 15 persons and it worked very well. There is no reason at all why it should not work equally well employing just 3 persons—one of whom must act as umpire and plot taker.

The greatest asset of this system is that it eliminates the need for the umpire to check each player's track lines and then try to relate them to his opponents. The rival track lines are interposed upon each other so that they can be compared at a glance.

Method

We use the graphic map again but this time the squares are used only as reference points. A useful scale to work on is 1 mm. equals 4 miles on the map and 100 ins. equals 20 miles on the floor (models).

The umpire calls the duration of the move which will be the period at which contact *could* occur. If contact was not possible for 5 hours, he would call a five hour move. It will be a big help to compile a speed chart showing the distance to move at any speed in a certain period of time. For example, using the scale suggested, a vessel moving at 19 knots will cover 23·75 mm. in five hours.

Speed (knots)	*1 hour*	*2 hours*	*3 hours*	*4 hours*	*5 hours*
1	·25	·5	·75	1	1·25
2	·50	1	1·5	2	2·50
3	·75	1·5	2·25	3	3·75
4	1	2	3	4	5
5	1·25	2·5	3·75	5	6·25
6	1·5	3	4·5	6	7·5
7	1·75	3·5	5·25	7	8·75
8	2	4	6	8	10
9	2·25	4·5	6·75	9	11·25
10	2·5	5	7·5	10	12·5

Track lines should be measured off and drawn lightly using a rule.

The table given below only covers up to 10 knots and five hours but you can easily extend the table to 12 hours and 35 knots by adding to the figures given here.

Map making

It is necessary to make two copies of the same map. One for each side. A third map, which is a reference tool for the umpire, is useful but not essential.

A transparent copy of the map is also required. It is not required to draw the whole map upon the transparency but only sufficient to enable it to line up perfectly on one of the originals. The two originals must be exact copies of each other. My own method was to draw one map and then staple the second blank graphic sheet to it (only in four places as it must be removed afterwards). Then I raised a portion of the whole up to the light so that the lines drawn on the lower map could be seen through the top sheet. I could then trace the map with complete accuracy.

The transparent top map can consist of large sheets of tracing paper. It has a rather limited life though, and has to be replaced at intervals.

What is much better is that material which looks like perspex but is in fact almost as pliable as paper. It is absolutely clear and similar 'foggy' material should be avoided. It is obtainable in your local do-it-yourself shop or ironmongers.

The size of the material required is 30 ins. × 22 ins. so that it will just overlap the 29 in. × 21 in. graphic sheet. The assistant will cut off sufficient for your purpose.

This is the ideal material to form your transparent copy. You can draw upon it in chinograph pencil and simply rub out obsolete lines as you go along.

Description of the method of play (See also diagrams on pages 116–118.)

1. Each player has a map but the second player (B) also has a transparent copy.
2. Player (A) draws his track lines directly upon the map in pencil.

3. Player (B) lines his transparent map directly atop the graphic map and draws his track lines upon the transparency in chinograph pencil.
4. When both players have finished extending the track lines of their forces, they hand in the maps to the umpire. He will lay the graphic map of (A) upon a table and line up the transparent map of (B) directly on top of it.

   Note. The graphic map of (B) stays behind. It is used only to display all the features of the map onto the transparency while he is making his calculations and drawing track lines.

It will be apparent that when the umpire (or plot-maker) lines up the two maps, it is as if he is looking at one master map with all the track lines displayed together for comparison. It is as if both players drew upon the same map. Now the umpire can see at a glance whether contact has occured between rival forces.

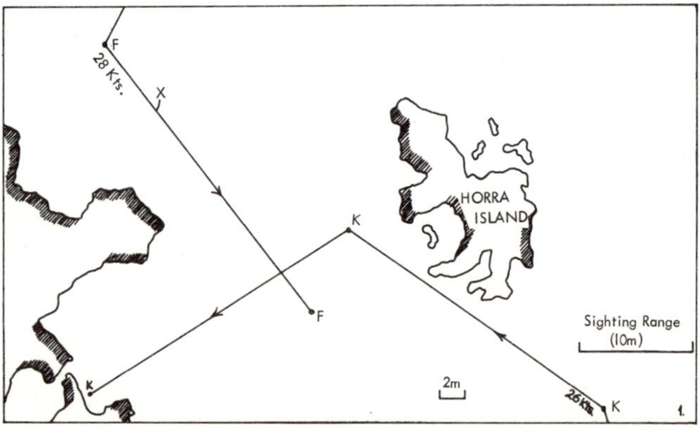

DIAGRAM (A)

The umpire sees that rival track-lines intersect each other. Has force (F) sighted (K) and vice versa ? The answer is no, since by simple calculation it can be seen that force (F) was in fact at point X when (K) intersected the line. The sighting range from X does not encompass the 'meeting' point.

The track lines may be single ships or a group of ships moving at the speed of the slowest (or at regular cruising speed). In the latter event, they must be coded. Formation plans of each force of ships must be drawn as explained by the diagrams 3 and 4 on page 110. The formation can be changed at any time if desired, provided that a fresh cruising plan is drawn. When contact has been made, the current plan is used in order to lay out the models.

Much of what has been written in all three methods of play described in this chapter can be included in any of the games. For example: the visibility figures suggested for the semi-secret campaign can also be applied to the completely secret game. All that I have done is to describe the thought behind each method. I have avoided a mass of detail and hard and fast rules because a wargamer seldom faithfully follows a vast set of rules. He prefers to mould and adapt a game to suit his own requirements and would undoubtably

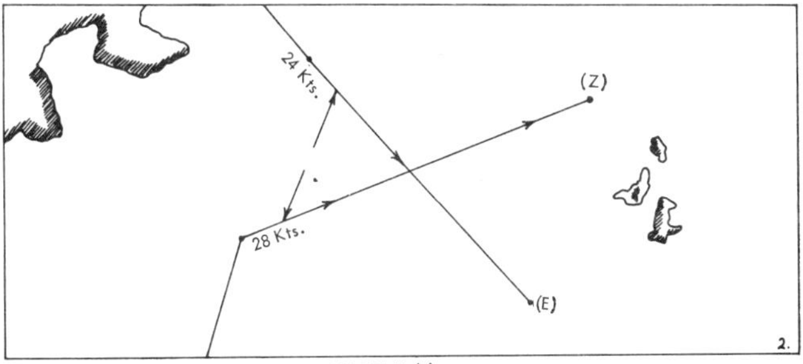

DIAGRAM (B)

Here a sighting has occured very shortly after the move started and therefore the track-lines beyond the sighting point are invalid and must be erased. The umpire or referee makes known that force (E) and (Z) have sighted each other and asks for the cruising plans of the rival forces.

FORCE (E)
Battleship "Gunfire"
H. Cruiser "Storm"
L. Cruiser "Thunder"
Destroyer "Sleek"
Destroyer "Swifter"

FORCE (Z)
L. Cruiser "Mallard"
L. Cruiser "Fearfull"

disagree with much of its contents: Wargamers are imaginative people who like to incorporate many of their own ideas. A lot of the fun of this hobby is in research and adapting or modifying other peoples ideas and including your own rules. Besides that, it would be unwise to present a complicated game with which the newcomer could not cope. Fight a simple action first, then when you are complete master of it you could decide to add radar (one example) to your game. There are no shortage of books dealing with the Pacific war. Read some of them—find out the facts about various types of radar and how effective they were and then make your rules.

To cover every facet of WW2 naval warfare would require a volume in itself and would not really help the wargamer because that which two persons can undertake is strictly limited. Whatever facet of modern warfare is considered, it is without doubt too involved to reproduce in full (unless we are fighting a very limited and specialised action). Submarine attack and surface attack on submerged submarines for example is a tremendously involved affair with an almost unlimited number of possible results. Moreover, the American procedure was always different and superior to the Japanese and depended on whatever radar and sonar was

The bearing of each force in relation to each other when the sighting occured is shown instantly by the map track lines.

With this knowledge and the known disposition of the forces from the plans (above), the models can be laid out on the floor.

It is assumed that the leading ship in each force sighted the enemy. Since 10 ins. on the floor equals 1 mile on the map and the two rival forces saw each other at the prevailing visibility of 10 miles, the models are laid out at the correct distance and bearing.

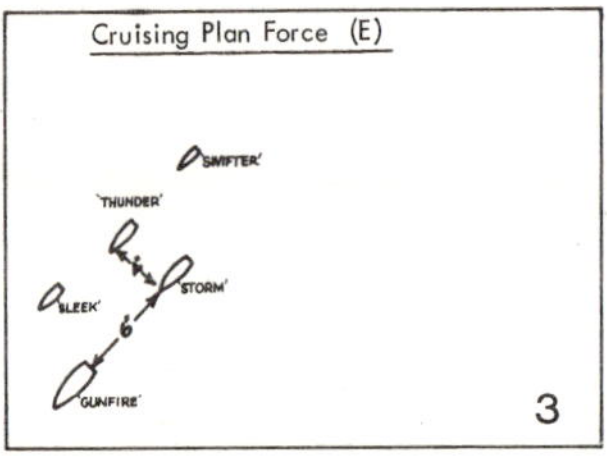

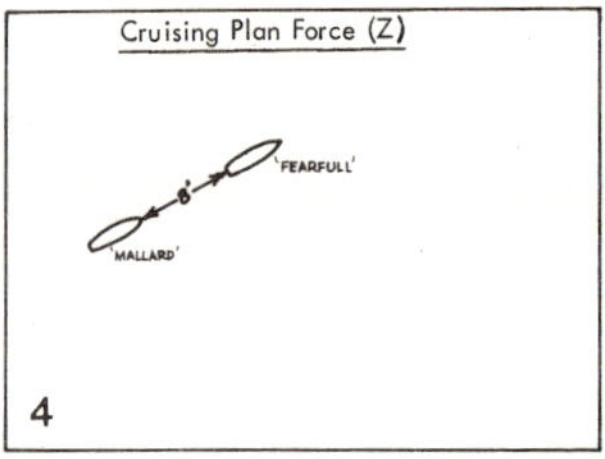

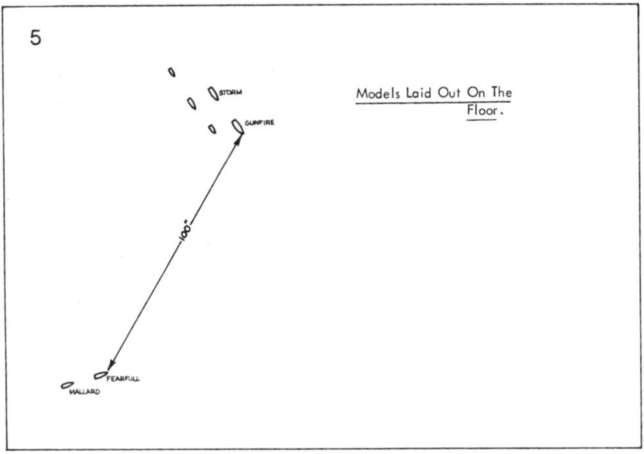

If we assume that gun ranges are as follows: 14-inch ('Gunfire'), 120 ins.; 8-inch ('storm'), 100 ins.; 6-inch (light cruisers), 75 ins.; we observe immediately that battleship 'Gunfire' can open fire upon the two light cruisers, but all other vessels are out of range.

When the next move is taken, the positions of the models and distances from each other are bound to alter.

at hand. This does not mean that our task is hopeless. We must devise a system that gives the principle of these attack systems and reproduces their broad outline.

A very simple attack system is given in the modern battle rules but these are intended for the first two types of campaign described in this chapter. Although all these campaign rules apply to a system of play rather than rules for a specific period, I have given them a 'modern' treatment. To this end I have included *below* rules for a more sophisticated method of submarine attack for use with the completely secret game. It is intended for WW2 operations. Submarine attack in the first world war was quite different. I have given these rules because most other WW2 fighting rules can be applied to this type of game but submarine attack needs specialised consideration. It is to be hoped that the rules will act as a pointer to the manner in which further rules can be drafted to suit any situation in any period.

SUBMARINE RULES FOR THE 'SECRET MOVE' CAMPAIGN

1. In order to fire torpedoes, the submarine must be within 2·5 mm. of the target (see diagrams A and B). This presumes the scale suggested is being used and represents 10 miles. This is the average sighting range of a surfaced submarine in daylight.
2. The umpire will ascertain whether the submarine is in a position to attack (according to track line bearings as in A & B diagrams) and if so, he will instruct the player commanding the targets to lay out models representing them according to his cruising plan.
3. The model submarine will be placed 10 miles (i.e. 50 ins. on suggested floor scale) from the target(s) strictly in accordance with the angles of bearing of the track lines on the map.
4. When the umpire is satisfied that all the models are laid out correctly, movement begins. Full moves are taken. One in. per knot at speed targets were steaming on the map (20 knots is 20 ins.)*

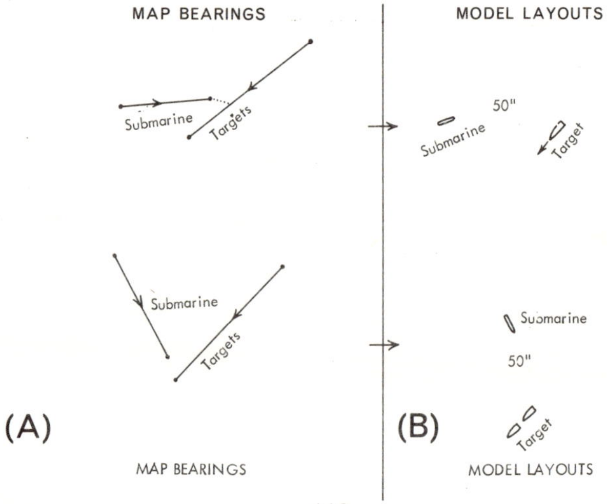

5. If the submarine is astern of the targets, or if they are steaming away from the direction of the approach, then it will be obvious that the submarine cannot intercept and attack. When such instances are obvious the umpire should not tell the player commanding the targets that he has been sighted. He must inform the submarine commander of the sighting however. If the umpire is in doubt, the models must be laid out and an attack attempted. If the submarine declines to attack in the first place models need not be laid out and targets commander should not be informed.
6. The submarine is deemed to be on the surface at the start of the move and may make a surface approach and dive at any stage of the approach. The umpire must be told one move in advance when the submarine is going to submerge. Submerged moves are at reduced speed. (Gato class subs 21 knots above and 10 knots submerged). Torpedoes may be fired when the submarine is on the surface or submerged. When the submarine is submerged the model is still used to indicate the underwater direction of the attack.
7. The submarine may be located either when submerged or on the surface. Targets commander throws I dice per move.
   Surfaced. 1.2.3 Targets locate submarine.
   Submerged. 1 Targets locate submarine.
8. When submarine is located, targets may take evasive action. Prior to this they must maintain course. All vessels present armed with depth charges may attack. If on the surface, gunfire can be directed at the submarine. Each vessel within range may fire once only and a throw of 1 is needed to destroy the submarine. If submarine is not hit, it is deemed to have submerged immediately afterwards (in normal gun actions the rules described on page 122 should be used).
9. All vessels capable of reaching submarine's position may drop depth charges (once per move). Using cards, a king must be drawn to denote destruction of the sub-

marine. After three moves, depth charging must cease.

10. All anti-submarine vessels have an asdic beam 6 ins. directly ahead of the bow.* Any submarine passing through the beam (in order to reach targets screened beyond) may be located. If 1.2.3 is thrown then the submarine is located and may be depth charged. No torpedoes may be fired when a submarine is located.
11. If not located the submarine may fire torpedoes not in excess of 24 ins. from any target. Torpedoes may be fired using markers or chance cards (see pages 130 and 132) but in order to compensate in part for unscaled ranges, only half the torpedoes normally fired may be discharged.† When torpedoes are fired the whole move is broken down into one-fifth short moves. This is to prevent torpedoes hitting on the first move and thus making missing impossible. Ships and torpedoes are moved at one-fifth normal speed. At 40 knots move at 8 ins. etc.
12. When a submarine is located other than by asdic, it may fire its torpedoes if in range. It must do so immediately however.

    It will be apparent that vital targets should have a destroyer screen and that these will need be 24-in. or greater distance on either beam of the targets to be effective. The screen can be fired at however if the targets cannot be reached or if the attack is to risky. If the screen is hit however, the submarine may then be depth-charged. If the submarine remains undetected the targets must maintain course and not attempt to avoid torpedoes. Submarines may not fire at a target at a range less than 15 ins. (reduce speed if necessary).

Torpedoes cannot hit the target in less than two moves. If the target is approaching bow-on and will 'run into' the torpedo in a single move, then the move will need be broken down into a smaller fraction.

*If scale 1-1200.

†Applies only to marker system.

I know that the whole thing appears vastly complicated but it is merely a question of following the sequence step by step and checking whether points mentioned apply to the situation at hand. Once it has been done, the second time is very much easier and soon most of it is committed to memory. Why not practise this method before incorporating it into a major game. Make a small map and stage a U-Boat attack on a convoy. Small actions like this can be good fun and excellent practice.

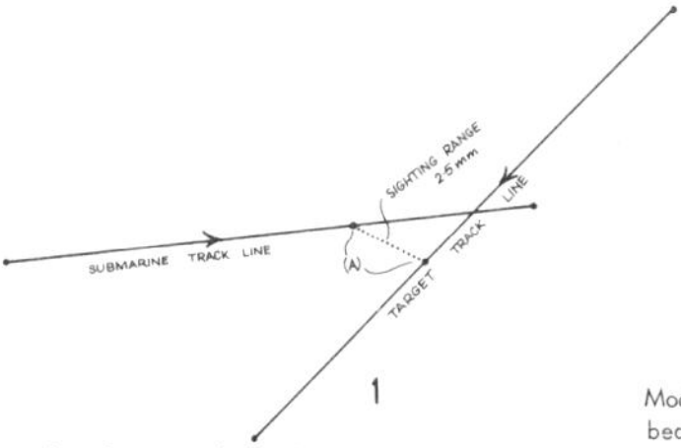

1

A = Point at which submarine sighted target.

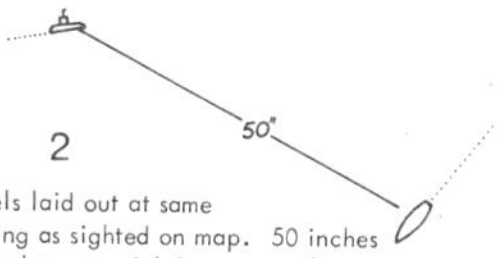

2

Models laid out at same bearing as sighted on map. 50 inches is equal to map sighting range of 2.5 mm.

Right hand drawings show how submarine and target is laid out according to map bearings on left.

(3) and (4) are bearings highly unfavourable for target interception.

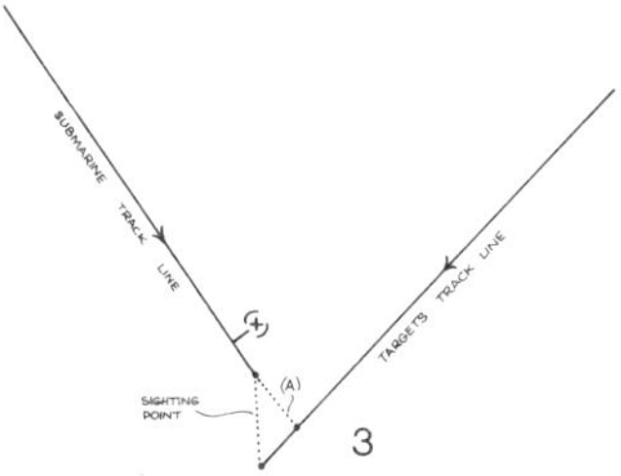

3

If it is asked why target was not sighted at 'A' the answer is that when target was there, submarine was at 'X' and thus out of sighting range.

4

Sam makes his map moves in pencil directly upon a drawn paper map.

The system given here is for daylight attack. Fight your first game on the assumption that night-time does not exist. Later on, you can include a period of night-time (say eight hours in twenty-four) and you will be kept very busy indeed formulating workable rules for fighting at night. You will also need find out all the facts about what happens when various forces and weapons are used in the dark. Confusion is heightened and friend very often mistaken for foe. Very often we can risk moving at night when we dare not venture out during daylight.

All that I have written is a method of play rather than an attempt to cover everything. In fact an awful lot is not covered at all simply because you *cannot* account for everything. It is intended that the reader will add to it as he desires. Let us mention a single aspect of the 'modern' game—torpedo attack. If you so desire, you could include rules for German circling torpedoes and acoustic torpedoes. If you research the latter, you discover the British counter to this menace—the squawker buoy and the German counter to that—the torpedo designed to steer ahead of the squawker buoy. I have already mentioned radar as another thing that can be included. There are a host of others. My purpose is to provide a broad method of play and leave it to you to make it suit your needs. The chapter as a whole simply tells how to stage map campaigns but the specialised submarine rules which are only usable with the completely secret rules are intended to show how special treatment can be given to any given subject.

Whatever you wish to include in your game, you will need to make the rules fit the method of play. This chapter describes three different ways that we can stage map campaigns and it will be necessary to make your rules and systems so that they work according to which one you are using. Rules that suit the self-contained map campaign would often need to be modified to fit the completely secret game. Time scales and things like that are entirely different.

If you merely stage a limited affair such as re-hunting the *Bismarck* then you can afford the luxury of formulating detailed rules. The author's opinion is that it is more rewarding to fight a miniature war and that the thrill of master minding grand strategy more than compensates for relatively simple rules.

Your task will be made easier for you if you decide to fight a Napoleonic campaign instead of a 'modern' one. A Napoleonic campaign *can* be made highly complicated of course but there is not nearly as much involved as in a game set in our present century.

2

Albert too has a paper map which is exactly the same as Sam's. Albert however places a transparent copy atop his map and draws his track lines in chinograph pencil on the transparent copy.

The only complete rules I have given are for fighting a miniature 'World War' campaign and here again the reader can adapt, modify, or add to it as he sees fit. The nature of this game meant that many approximations had to be made that might offend the sticklers for realism. Those who may prefer to stage a more limited and sophisticated game, will have the ideas to do so I hope, as a result of reading these pages.

If you are able to form a group of battlegamers, then your game can become more involved and realistic because there are more people to handle the complication. Some members can specialise so that it is not necessary for everyone to know all the rules. The umpires can be armed with the volume of rules and inform the players of facts and figures when the need arises. If for instance, one player wished to put his crippled carrier into port for repairs, he need not know all the rules in advance. When the matter concerned him, he could either look up the relevant data or consult the specialist in this field.

The more people taking part in your game or club the grander and more sophisticated it can become. If it is only practical for one person to command a maximum of 10 ships, then 20 players in the group could command between them a force of 200 ships.

EDGE OF WALL MAP SHOWING SQUARE LOCATION CODE

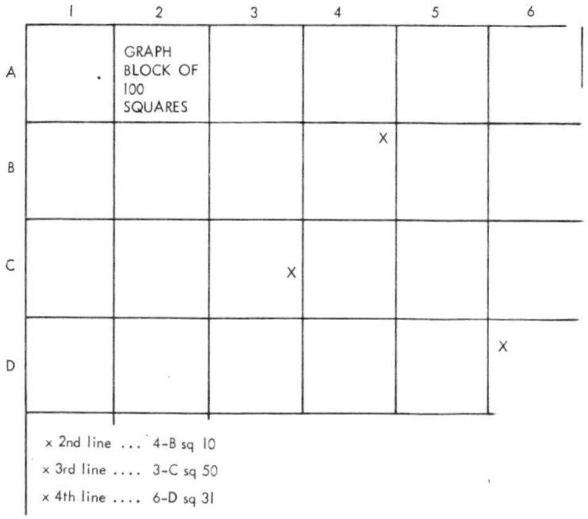

Page from your fleet list.

Aircraft carrier No. 1, *Thunderblast*; 'Fourwind' class.
23,709 tons, 2-inch armoured deck, 31 kts., 70 aircraft.

Guns		*Damage*
12 5-inch A.A.	1,200	
30 40-mm.	850	
120 25-mm.	2,050	
	7,320	
	120	
	9,490	

Aircraft aboard	*Aircraft losses*	*Replacements*
15 divebombers	1.3.2	6
20 torpedo-bombers	7	7
35 fighters	1.2.2.1	

Enemy ships sunk	*Enemy aircraft destroyed*
Battleship *Storm*	15.3.8
Submarine *Swordfish*	

The lower section (enemy casualties) is non essential data!

CHAPTER 8

World War II Fighting Rules

Although intended primarily as a battle system for second world war type fighting, these rules and systems are quite adaptable for all the years going back to around 1880. Large pre-dreadnought fleets can use these rules with most modifications necessary only applicable to rates of fire and ranges. Rules for aircraft carriers and submarines are simply ignored.

The rules should be regarded as a basic system to which some modifications will need be made depending on the circumstances. If the rules are applied to the self contained map game then needs will differ slightly than when the reader uses models to fight his battles.

The rules may appear complicated at first. In fact they are not. It is entirely practical to fight fairly large actions using this system. In truth, as far as WW2 goes, our concern is to be able to apply fighting rules to the 'World War' game described elsewhere. As we can see, some really major actions are likely to take place from time to time. These rules have stood the test of time. If your need is only to fight small battles, then elaborate them as you desire to fit your needs.

SIMPLE BATTLE DAMAGE POINTS SYSTEM

Simple, yes! The system is just as effective though as many complicated formulas that I have seen. What is it that defines how difficult it is to sink a vessel?

1. Her size (tonnage).	There are other factors—such as
2. Her protection.	the quality of her sub-division etc.
3. How well built she is.	but I tend to include such things under (3).

I account for these factors simply by awarding 1 point per ton (standard displacement) and 1,000 points per in. of armoured belt at its thickest point. Let us take the well known *Bismarck* as an example. She is 42,000 tons and has an armoured belt of 12·6 ins.

thus: 42,000 points for tonnage
12,600 points for armour

54,600 p.v. total.

We do know that German ships were superbly built and very difficult to sink. I always add 33 per cent of total points value to German ships.

Weight of shell and rate of fire are both used to compute damage caused by gunfire. It is easiest to standardise on both so that we give a 12 in. gun a rate of fire of two rounds per minute and a shell of 850 lbs. Note how the latter can vary though. Actual shell weights of each ship (when known) may be used if desired by the battlegamer.

Italian	997 lbs.
German	980 lbs.
French	950 lbs.
U.S.A.	870 lbs.
British	850 lbs.
Russian	714 lbs.

WW1 figures. The American 12 in. gun of WW2 fired a shot weighing 1,140 lbs.

Using weight of shot is useful because it reflects the increased damage shells cause as their calibre increases inch by inch. Thus we see a 11 in. gun firing a 660 lb. shell will cause nearly 7 times as much damage as a 6 in. gun firing a 100 lb. projectile. How accurate is this?

Let us take a 2,000-ton destroyer as an example. We can see that three or four 12-in. shells will sink our destroyer and that it will take twenty 6-in. shots to do the same job. As an approximation, this is reasonably realistic. We can easily vary the damage caused by successive hits by the use of tables, though this must add complication and time to our game.

We all know that two hits by the same calibre shot may cause vastly different damage. On the other hand, damage from a whole series of hits would probably average out and reflect upon the size of the shells in question.

Whenever a shell hits the target, we simply convert the weight of the shot in pounds into points valve. A 100-lb. shell thus causes 100 p.v. damage.

USING THE GUNNERY CHART

The gunnery chart looks rather alarming at first glance but its use in practise is remarkably easy.

This is the procedure.

Having measured the range to the target simply trace the range figure opposite the calibre of the gun firing. All that is necessary is to note which card is listed right at the top of the column where you found the range figure.

Here is an example:

Gun firing 14-in. Range to target 71 ins.

We find the 14-in. gun listed in the first column. Opposite this we quickly find 71-in. in the 5th column and hence under the card 5.

What this means is that we must cut a card up to and including No 5 and of course it must be a *black card* (Red cards are always misses). Any card higher than 5, regardless of its colour is a complete miss.

It will therefore be observed that as the range closes so the chances of hitting the target increases.

If the card was *black* and within the group specified to hit, we now consult the second table for number of hits.

We use the *same* card because the number of hits will depend upon it and if we were firing at long range then we needed a low card in order to secure any hits. A low card on the second table will not permit us to score a large number of hits. This ties in with the fact that the range was great and that at long range large numbers of hits are not possible.

The procedure is to simply locate the number of guns that were fired at the target and opposite that under the column headed by the card that you have drawn will be listed the number of hits scored. Thus if you fired 9 guns and your card is a 5, then 2 hits have been scored. Had 12 guns been fired, then 3 hits would have been obtained. With a little experience the whole procedure is easy and rapid.

GUNNERY INFORMATION

Calibre	*Shell weight*	*Rate of fire*	*Weighted for r.o.f.*
18 in.	3,200 lbs.	2	3,200
16 in.	2,500 lbs.	2	2,500
15 in.	1,900 lbs.	2	1,900
14 in.	1,400 lbs.	2	1,400
13·5 in.	1,250 lbs.	2	1,250
13 in.	1,200 lbs.	2	1,200
12·6 in.	1,157 lbs.	2	1,157
12 in.	850 lbs.	2	850
11 in.	660 lbs.	2	660
10 in.	500 lbs.	3	625
9·2 in.	380 lbs.	4	570
8 in.	275 lbs.	5	550
7·5 in.	200 lbs.	5	500
7 in.	165 lbs.	6	495
6·5 in.	115 lbs.	7	475
6 in.	100 lbs.	8	450
5·5 in.	84 lbs.	9	405
5 in.	60 lbs.	10	360
4·7 in.	45 lbs.	10	336
4 in.	31 lbs.	12	245
3·5 in.	25 lbs.	14	200
3 in.	13 lbs.	15	140

Use column 4 as individual shell damage figures for the easiest way in which gunnery can be dealt with. All ships fire one salvo and the damage shown in column 4 is multiplied by the number of hits scored.

GUN RANGES AND CARD REQUIRED TO HIT (BLACK CARDS ONLY); WW2 ACTIONS

Long-range Chart

Gun	*Ace*	*2*	*3*	*4*	*5*	*6*
18 in.	92-87	86-83	82-79	78-75	74-71	70-67
16 in.	88-85	84-81	80-77	76-73	72-69	68-65
15 in.	86-83	82-79	78-75	74-71	70-67	66-63
14 in.	84-81	80-77	76-73	72-69	68-65	64-61
13·4 in.	83-80	79-76	75-72	71-68	67-64	63-60
12·6 in.	80-77	76-73	72-69	68-65	64-61	60-57
12 in.	78-75	74-71	70-67	66-63	62-59	58-55
11 in.	76-73	72-69	68-65	64-61	60-57	56-53
8 in.	70-67	66-63	62-59	58-55	54-51	50-47
6 in.	67-64	63-60	59-56	55-52	51-48	47-44
5·5 in.	66-63	62-59	58-55	54-51	50-47	46-43
5 in.	65-62	61-58	57-54	53-50	49-46	45-42
4·5 in., 4·7 in.	63-61	60-57	56-53	52-49	48-45	44-41
4 in.	63-60	59-56	55-52	51-48	47-44	43-40
Smaller	60-57	56-53	52-49	48-45	44-41	40-37

Short-range Chart

Gun	*7*	*8*	*9*	*10*	*Jack*	*Queen*
18 in.	66-63	62-59	58-55	54-51	50-47	46-43
16 in.	64-61	60-57	56-53	52-49	48-45	44-41
15 in.	62-59	58-55	54-51	50-47	46-43	42-39
14 in.	60-57	56-53	52-49	48-45	44-41	40-37
13.4 in.	59-56	55-52	51-48	47-44	43-40	39-36
12·6 in.	56-53	52-49	48-45	44-41	40-37	36-33
12 in.	54-51	50-47	46-43	42-39	38-35	34-31
11 in.	52-49	48-45	44-41	40-37	36-33	32-29
8 in.	46-43	42-39	38-35	34-31	30-27	26-23
6 in.	43-40	39-36	35-32	31-28	27-24	23-20
5·5 in.	42-39	38-35	34-31	30-27	26-23	22-19
5 in.	41-38	37-34	33-30	29-26	25-22	21-18
4·7 in., 4·5 in.	40-37	36-33	32-29	28-25	24-21	20-17
4 in.	39-36	35-32	31-28	27-24	23-20	19-16
Smaller	36-33	32-29	28-25	24-21	20-17	16-13

GUNNERY TABLES

(please check your card spade/club)

Black Cards Only

Number of guns firing		*A*	*2*	*3*	*4*	*5*	*6*	*7*	*8*	*9*	*10*	*Jk*	*Qu*	*Kg*
15	Clubs	1	2	3	4	5	6	7	8	9	10	11	12	13
	Spades	3	4	5	6	7	8	9	10	11	12	13	14	15
14	Clubs	1	1	2	3	4	5	6	7	8	9	10	11	12
	Spades	3	3	4	5	6	7	8	9	10	11	12	13	14
12	Clubs	1	1	1	2	3	4	5	6	7	8	9	10	11
	Spades	2	3	3	4	5	6	7	8	9	10	11	12	1
10	Clubs	1	1	1	1	2	3	4	5	6	7	8	9	10
	Spades	2	2	3	4	4	5	6	7	8	9	10	1	5
9	Clubs	1	1	1	1	2	2	3	4	5	6	7	8	9
	Spades	1	2	3	3	4	4	5	6	7	8	9	1	4
8	Clubs	1	1	1	1	2	2	3	3	4	5	6	7	8
	Spades	1	2	3	3	3	4	5	5	6	7	8	1	3
7	Clubs	1	1	1	1	2	2	2	3	3	4	5	6	7
	Spades	1	1	2	2	3	3	4	4	5	6	7	1	2
6	Clubs	1	1	1	1	1	2	2	2	3	3	4	5	6
	Spades	1	1	2	2	2	3	3	4	4	5	5	6	6
5	Clubs	1	1	1	1	1	1	2	2	2	3	3	4	5
	Spades	1	1	1	2	2	2	3	3	3	4	4	5	5
4	Clubs	x	1	1	1	1	1	1	2	2	2	3	3	4
	Spades	x	x	1	1	2	2	2	3	3	3	4	4	4
3	Clubs	x	x	1	1	1	1	1	1	2	2	2	3	3
	Spades	x	x	x	1	1	1	1	2	2	2	3	3	3
2	Clubs	x	x	x	1	1	1	1	1	1	1	2	2	2
	Spades	x	x	x	x	1	1	1	1	1	1	2	2	2
1	Clubs	x	x	x	x	1	1	1	1	1	1	1	1	1
	Spades	x	x	x	x	x	x	x	1	1	1	1	1	1

This system is very quick in operation and ideal for large battles. When circumstances dictate, I prefer to account for individual shots hitting so that the number of hits obtained might vary from one right up to the number fired multiplied by rate of fire.

For example, if we are firing nine 4-in. guns, we could score any number of hits from one to 108 since the gun fires twelve times per minute. When the shot is weighted for rate of fire, we cannot score more than nine.

Obviously, more work is required to determine any number of individual hits but here is how it is done.

Draw your card to determine number of hits scored as already directed. However you now multiply hits scored by rate of fire of gun to arrive at a figure that represents the maximum possible number of hits that you could score. You then cut another card to see the *actual* percentage of this figure that you score.

Suppose your ten gun ship scored five hits with 8-in. shells. You multiply this by rate of fire. R.O.F. of 5 r.p.m. × 5 hits per salvo equals maximum possible hits of 25.

Now presume you cut a six card using the table below. This indicates 30 per cent of your 25 hits are counted. We cannot get an exact figure but eight is the nearest and so eight hits are scored. If we had the figure 108 hits instead of 25 and cut a Jack, then 88 hits would be obtained. When this system is used, only count actual weight of shot (col 2 instead of col 4) when computing damage.

Ace to three	5%		
4	10%	9	60%
5	20%	10	70%
6	30%	JK	80%
7	40%	QU	90%
8	50%	KG	All.

When this latter method of determining hits is used a normal firing move should be missed because of the extra fire-power available. Fire every 2 minutes instead of every 1 minute. Movement occurs in 20 second periods and firing normally takes place after 3 moves or each minute. See movement rules.

Range Figures

The range figures given in the gunnery chart may be too high for the space available. Instead of counting the figures as inches it may be more practical to count them as half or quarter ins. It will depend to a large extent on the scale of your models. When applied to battles that take place on a map, these figures can represent squares or millimetres or any other measurement. Or you can compile your own chart and substitute the figures given for your own.

Movement of Model Ships

Presuming the use of the ever popular 1-1200 scale model, a scaled speed of 28 knots is simply 28 ins. (Per minute). Each minute move is sub-divided into three phase moves. So each phase-move is just one-third of the warship's actual speed in ins. Firing takes place every third phase move or after one minute. The reason for the phase move is so that enemy tactics can be more readily countered.

The amount of damage we can do according to these rules, is vastly in excess of what a real ship could do in one minute! The relationship between moving and firing *must* be compromised otherwise 20 minutes' firing would involve moving vast distances at scaled speed.

Therefore the minute move must represent about 20 minutes' firing or more but our main concern is to reproduce the effect of fighting a battle rather than struggle to do it all absolutely correctly. It is possible only if tiny scale models are used.

Firing may occur every phase move or every second phase move as directed in the rules, if a quicker decision is desired.

Firing procedure

1. Measure range to target (bow to bow).
2. Observe number of guns that can bear upon the target. Mark down on a firing pad the target and number and calibre of guns firing at her. When both sides have written down all the targets, draw cards for effect in turn and under observation of your opponent.

	hits	*damage*
Bismarck at *Hood*	4-15 in.	
Bismarck at *P.O.W.*	4-15 in.	
Bismarck at *L.Cru*	8-6 in.	

Any set of guns may not fire at more than two targets.

If the line of sight between any warship and its target is 'spoilt' by the presence of another friendly ship then the following rules are to be observed.

(A) A warship cannot fire over another of the same type or larger.

(B) It is permissible to fire over a smaller type provided it is 12 ins. away or more.

(C) If the target is two types or categories smaller it is permissible to fire over it from 4 ins.

Destroyers are not allowed to fire at battleships.

Turning

This could be made very complicated because it depends upon so many factors. I shall introduce a simple but satisfactory system. If individual players wish to make it more comprehensive and involved they are free to do so.

Battleships	45 degree turn	These are maximum permitted turns. Only one turn may be made per phase move.
Cruisers	90 degree turn	
Smaller	180 degree turn	

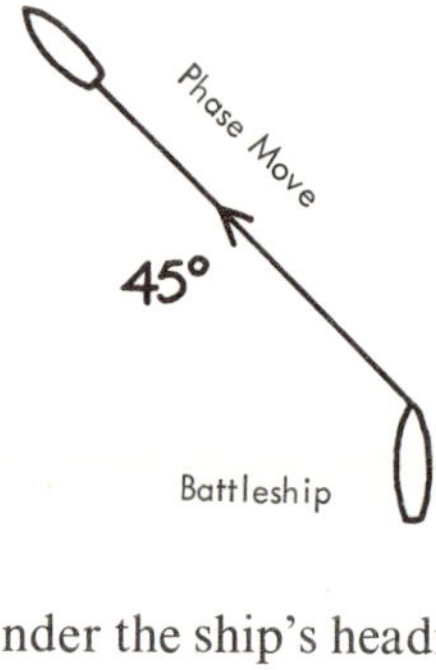

Damage

Record damage under the ship's heading in your fleet list for which space should be provided.

(A) If a vessel has armour then her armour points value plus half of her tonnage points value must be lost before she is crippled and reduced to half speed and half gun power (i.e.) 5 guns instead of 10.
(B) If the vessel has no armour then she is crippled when half of her tonnage points have been lost.
(C) Carriers and Transports suffer twice the damage of other warships.
(D) Every time an aircraft carrier loses 20 per cent of its points value a card is drawn to see whether an explosion occurs.

Ace to 6	No explosion.	Ordinary explosion
7 to 9	Explosion.	25% points value lost.
10 to Queen	Major explosion.	Major explosion
King	Total destruction.	50% points value lost.
		(additional to that lost already)

(E) If aircraft are aboard a carrier at the time damage is inflicted a like proportion of those carried to damage percentage suffered are destroyed.

Anti-aircraft fire

You should specify the distance at which nearby warships can lend A.A. support to the vessel attacked by aircraft according to the size and scale of the map or models in use. For 1-1200 models a distance of 12 ins. is reasonable.

Draw 1 card for each ship able to fire A.A. guns.

	Battleship and carrier	*Cruiser*	*Destroyer*	*Smaller*
Ace	5	4	2	1
2	5	4	2	1
3	4	3	2	
4	4	3	1	
5	3	2	1	
6	3	2	1	
7	2	1		
8	2	1		
9	1			
10	1			

Cards not listed indicate none destroyed.
Spades only: double kills shown.

This system is perfectly adequate for the average wargame. To those who wish to account for the actual number of A.A. guns of each type aboard ship it is quite easy to make charts based upon the points value of the total guns carried, i.e. award 1 point for each gun or nest of guns. The higher the p.v. of guns carried the more likelyhood of bringing down a larger number of aircraft. A word of warning here. Do not make A.A. guns too effective.

Only aircraft that have survived fighter attack (if any) and then A.A. fire, may then draw cards for bomb hits.

Surface torpedo attack

This *must* occur after gunnery so that only those surviving any gunfire may fire torpedoes. Any surface vessel suffering 50 per cent damage (i.e. crippled—see Damage rules) may only fire half her tubes that are able to bear upon the target.

Warships carrying reloads of torpedoes may not reload until 4 phase moves later and then only if not under fire. Compatible rules for tactical battle maps should be in force.

Other vessels must return to port or base for reloads.

Torpedoes must be moved at scaled speed exactly similar to surface ships. They do not start to move until the following ship move is over.

Two systems can be used. The first is only suitable when models are used. It may at first sight appear somewhat crude, but in fact it does give very good results. Great care must be taken however.

Torpedoes are represented by match sticks or something similar. At the base is marked the setting unless one match represents a group of torpedoes when the setting of individual torpedoes should be noted on paper. The important thing is not to knock or disturb the matches and to ensure that they exactly maintain their course.

Of course it is easy to avoid a match as it approaches toward a target but the match is merely a marker for the actual position of the torpedo or torpedoes that it represents. A reference indicates the true position.

i.e. Marker No. (5) 1 torpedo 6½ ins. to left. 1 torpedo

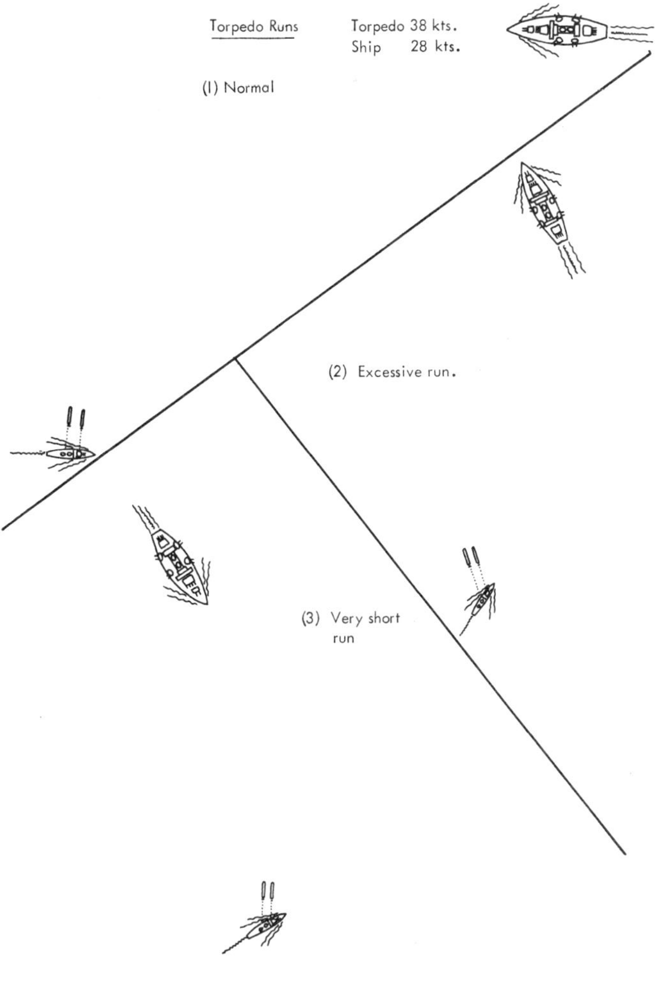
Torpedo Runs
Torpedo 38 kts.
Ship 28 kts.
(I) Normal
(2) Excessive run.
(3) Very short run

19 ins. to right. 1 torpedo 20 ins. to right.

Whenever it is suspected that a torpedo may have hit a ship, measurement occurs. The torpedo plot must be shown to the opponent to prove that a hit or hits have been scored. A marker may represent any number of torpedoes from 1 right up to the whole salvo fired. Futher, a number of torpedoes may be marked as running together.

In order to ensure that the marker proceeds exactly on course, a long straight edge or steel rule should be used.

What has to be borne in mind is that when gun scales have to be reduced (because of lack of space) then the ships are too close and thus too easy to hit. Therefore:

(a) Reduce number of torpedoes that can be fired.
(b) Allow only portion of ship to be hit. Say one-third of hull.

Depth Setting

Shallow: If set shallow and hits Battleship, only half damage counted.
Deep: Will run under a destroyer or smaller vessel.
Ranges: Vary greatly according to period and model of torpedo. Some ranges are given on the torpedo effect chart on page 139.

Torpedo hits by chance cards

This system is for map games. When fighting is conducted on maps, it is necessary to use a method of chance to decide whether hits have been scored. The odds of scoring hits however, must depend upon certain factors. The method given here can be simplified or elaborated as desired.

As an approximation, the average torpedo will assume to have a range of half that of a 14-in. gun. This table is therefore based upon the gunnery chart in this chapter and maximum torpedo range is thus 42 ins. (Or sqs). The odds of obtaining hits are computed according to the range, number and size of targets. Points are awarded according to the tactical situation. The first step is to state which ship is the specific target. The next step is to count the ships within a specified

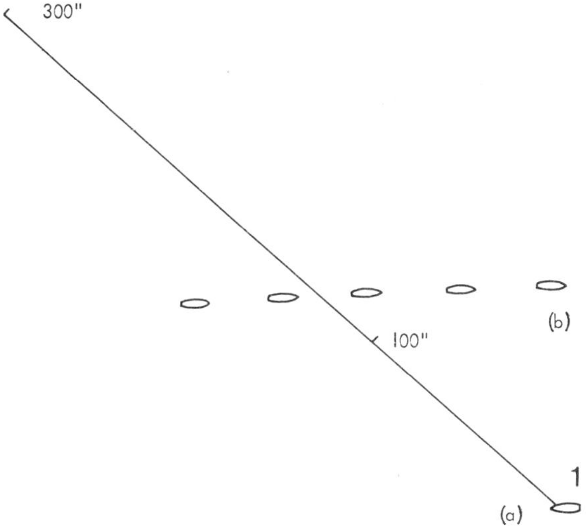

Solving the problem of scale: Diagram (1) shows what is presumed to be the scaled gun range of the ship (*a*). This gun range is assumed to be 300 ins., which is a distance too far to be practical. The range is reduced to 100 ins., but now we find that only the lead two ships in line (*b*) can be brought under fire.

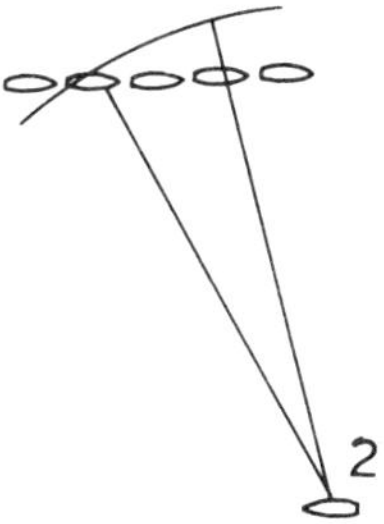

To help correct this situation, the models do not keep proper station (two lengths apart) but are allowed to steam very close. Now four ships can be brought under fire. The situation is partly restored.

Now it will be obvious that the models are too easy to hit by torpedoes and so only the first third of each ship is counted as far as torpedo hits are concerned. Those striking aft of this portion are misses and keep running.

area of the target. Say 5 squares if using the self-contained map described elsewhere.

Award points as follows:

Battleships and Carriers	6 points.
Cruisers & Merchant ships	4 points.
Destroyers	2 points.
Corvettes, Frigates, etc.	1 point.

If targets are heading bow on towards the direction of attack, then halve the total points awarded above.

The next procedure is to award 1 point per square less than maximum torpedo range at which the destroyer is firing. Example: Maximum range is 42 ins. but destroyer firing from 30 ins. (squares when related to map). Thus 12 points are awarded.

Having reached a points total, we use each 4 scored to 'climb' one card on the table below which is used to determine hits. A score of less than 4 means that only the first card on the table (Red Ace) will indicate a hit. Cut one card per torpedo fired. All other cards would be misses. However, if you had a points score of 36, then all Red cards from Ace to nine would indicate a hit. If the score was 56, then *all* Red cards plus Black Ace would be a hit.

Torpedoes should not be fired at ships heading *away* from the path of the torpedo at an angle greater than the diagram indicates.

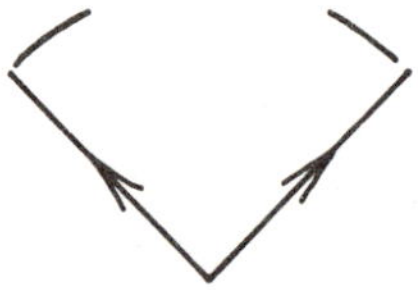

X Firing Point

When a hit or hits are obtained, players may if they wish, further dice or cut cards to decide which ship(s) in the group within 5 squares of specified target actually recieve each hit. Otherwise, allow target to be hit.

It is *not* necessary to deduct numbers of torpedoes fired because of errors in scale. This applies only to the use of models.

Red cards	*Score*	*Black cards*	*Score*
Ace	4	Ace	56
2	8	2	60
3	12	3	64
4	16	4	
5	20	5	
6	24	6	
7	28	7	
8	32	8	
9	36	9	
10	40	10	
Jack	44	Jack	
Queen	48	Queen	
King	52	King	

When score takes cards into *Black* list, all red cards are a hit.

Method

Destroyer attack occurs after gunnery. Obviously, destroyers moved into close proximity of targets (to increase chances of getting torpedo hits) will likewise be easier to hit by surface ship gunfire. No attempt is made to 'move' a torpedo. Those destroyers that survive gunfire simply select the target and compute the points according to the tactical situation and then draw cards (one for each torpedo) to determine hits scored (if any).

Minefields

These rules are only suitable for map campaigns. It is hardly practical to attempt mining operations when models manoeuvre upon the floor. After all, the largest hall represents only a scrap of ocean. The map on the other hand, represents a vast area and it is on the map that we must lay our minefields.

100 mines must be laid in order to effectively mine one graph square. Only two squares may be mined per move per ship. If moves are in factors of hours rather than a whole day, then specify one square mined in 6 hours. Mines may not be laid in excess of 150 miles from a coastline.

Minesweepers Can sweep one square clear per day.
Hitting Mines You should note the square numbers mined. When an enemy vessel or group of vessels passes over the mined square(s) cards must be drawn for:

(a) Odds of vessel hitting a mine.
(b) Effect of mine hit (if any).

Draw one card for each vessel that has steamed over the minefield during the move.

Black Card Vessel strikes mine.
Red Card Vessel steams clear.

If a vessel strikes a mine, use the 24-in. torpedo chart to see the damage caused.

Smoke screens (for use with models).

1. Destroyers only will be permitted to lay smoke.
2. Not more than $\frac{1}{4}$ of destroyers present may lay smoke screens at any one time.
3. Smoke is represented by laying cotton wool the distance travelled by the destroyer during the move (or shorter if desired).
4. Smoke already laid blows clear upon the following move.
5. Ships may fire into the smoke screen. Four cards lower than that normally needed to hit must be counted and if hits result, then only one half are counted.

SUBMARINE RULES

A fairly detailed set of rules for use with a combination of models and the completely secret map campaign are to be found on page 112.

The rules given here are for use with the other two types of graphic map campaigns described in this book. These games are the 'Self Contained' & 'Semi-Secret' campaigns in which flags are used to represent shipping. The rules are in keeping with the concept of those games which are designed for the use of large fleets.

1. Submarines moving on the surface must be represented by a flag as are other vessels.
2. When submerged, remove the flag and determine which

square the submarine can reach at submerged speed. Underneath the submarine heading in your fleet list (devote a special portion of the page) write down the exact location of the submarine and cross out its former position. A reference may read E 19 Square 25. This means the graph block where E and 19 interpose (see page 119) and the square number indicates the exact square within the graph block.

3. Submarines are moved after all other surface ships have been moved. If there are no targets within range, or if it not desired to fire at targets that *are* in range, move the submarine as in (2) after finding position. If the submarine fires at targets, it must not be moved until the following move (if it survives attack).
4. Firing range is 8 squares. If the enemy flag is coded as in semi-secret rules, then the enemy must disclose all the ships represented by that flag so that the submarine may pick target. Each submarine may fire at two separate targets within range, i.e. if firing 6 torpedoes, 4 could be fired at one ship and 2 at another.
5. For each torpedo fired, cut one card. Ace to 5 is a hit.
6. If a submarine fired at one of your ships, you know within an 8 square radius where it must be. If it fired at two ships, the area is narrowed down according to the position of the two ships (does not apply if a single code flag attacked). Destroyers may search this area with asdic if they can reach that position. This occurs upon the following move.
7. To simplify the procedure, two destroyers only search for the submarine. More than two is not allowed. The beam from each is as follows. Any 10×8 square (drawn lightly in pencil) in any direction in 'front' of the destroyers own position. The submarine commander checks his secret position. If located in one of the two beams he must report 'contact made'.
8. The destroyer that located the submarine may then depth-charge any 45 of the 80 squares within the beam. Make very light dots in pencil. If the submarine position

is on one of the marked squares it is sunk. If the submarine is not located—or located but not sunk, the secret reference must be shown to the opponent to prove that it was not located. If not sunk, the submarine may carry on moving as before.

TORPEDO EFFECT CHARTS

World War 1

	17·7 in.	*18 in.*	*19·7 in.*	*21 in.*	*23·6 in.*
Ace	350	400	500	550	700
2	850	925	1,075	1,150	1,375
3	1,200	1,310	1,535	1,645	1,980
4	2,000	2,165	2,500	2,665	2,480
5	3,000	3,245	3,750	3,995	4,725
6	4,000	4,245	4,750	4,995	5,725
7	Dud	Dud	Dud	Dud	Dud
8	5,000	5,365	6,115	6,480	7,605
9	6,000	6,550	7,675	8,225	9,910
10	7,000	7,550	8,675	9,225	10,910
Jack	8,000	8,825	10,510	11,335	13,860
Queen	9,000	9,825	11,510	12,335	14,860
King	10,000	10,825	12,510	13,335	15,860

For mine effect, use 23·6 in. line.

World War 2

21 IN. TORPEDO EFFECT CHART

	Hearts	*Clubs*	*Diamonds*	*Spades*
Ace	Dud	Dud	Dud	Dud
2	500	750	900	1,000
3	1,000	900	750	500
4	1,500	2,000	3,000	4,000
5	5,000	4,500	3,500	2,500
6	Dud	6,000	5,500	7,000
7	6,500	Dud	8,000	9,000
8	11,000	10,000	9,500	7,500
9	8,500	12,000	13,000	100
10	350	14,000	12,500	Dud
Jack	15,000	15,500	2,000	25,000
Queen	16,000	4,000	18,000	5,000
King	6,000	7,000	8,000	10,000

For 24 in. torpedo and mines draw another card, unless a *dud* has already been drawn. Hearts add 100, spades add 1,000, clubs add 2,500, diamonds add 5,000.

TORPEDO RANGES

	World War 1		*World War 2*
23.6 in.	13,000 yds. at 27 kts.	21 in.	15,000 yds. at 28 kts.
21 in.	10,000 yds. at 29 kts.		10,000 yds. at 34 kts.
19·7 in.	10,000 yds. at 27 kts.		6,000 yds. at 46 kts.
18 in.	8,000 yds. at 26 kts.	24 in.	44,000 yds. at 36 kts.
17·7 in.	8,000 yds. at 24 kts.		22,000 yds. at 49 kts.

Commerce raiders

Instead of representing the raider with a flag, it is moved in the same manner as a submerged submarine (its position plotted). The raider has the advantage of high speed but it is subject to the sighting range of surface ships. When the raider is due to be moved its nature and position must be revealed if an enemy vessel(s) has moved into sighting range of it. All vessels able to reach the raider in one move (i.e. move and fire upon it) may do so. The raider may also move. All vessels that move to attack the raider are in effect, taking their next move in advance. Since the raider has not already moved it need not forego the next move.

The purpose of the commerce raider is to attack unescorted merchant ships and not seek action with warships. Armed merchant cruisers have sighting range reduced by one-third to simulate the fact that they are difficult to identify.

AIRCRAFT RULES

These rules are intended for use with the 'world war' map campaign. No symbols are necessary for aircraft. A record of each squadron's movement is kept (i.e. flew from Pralle airfield to Passen airfield on 7.11.40) so that an attack can be launched from any airfield knowing what aircraft comprise the flight. The target is stated. The defender musters his fighter defence (if any) and the aerial battle is staged.

Those bombers that survive the encounter then go on to attack the target be it shipping or land targets.

The range of aircraft related to the World War campaign is as follows.

Heavy Bombers	220 squares
Other Bombers	110 squares
Fighter Bombers	85 squares
Fighters	60 squares

Defending fighters must be able to reach the target in order to attack the bombers. Here is the sequence.

1. Fighters attack escorting fighters. Air battle between the fighters.
2. Surviving fighters then attack the bombers. Bombers shoot at attacking fighters. Cards are drawn and losses computed according to number of aircraft present.
3. Surviving bombers then endure A.A. fire if any. Survivors of A.A. fire then attack target and draw cards for bomb effect.
4. All losses are noted on paper as they happen and after the battle the squadron losses are clearly noted in your squadron record. If five squadrons suffered heavy losses, it is possible to reform say three squadrons utilising the total aircraft remaining.

FIGHTER COMBAT

Both sides cut cards for the total number of aircraft present at the *beginning* of the battle.

Per 50 aircraft	*Per 10 aircraft*	*Individual aircraft*	
Hearts: Kills as on card. (Jack is 11).	Red: Nil	Ace–2	1 kill
Spades: Double kills shown on card.	Black: As on card	3	2 kills
		Others	Nil

Diamonds:

Ace to 5	Treble kills
6	28 kills
7	30 kills
8	31 kills
9	32 kills
10	33 kills

If you attacked with 62 fighters:

(*a*) Cut 1 card for 50;
(*b*) Cut 1 card for 10;
(*c*) Cut 2 cards for individual combat.
Total 62.

Jack	34 kills
Queen	35 kills
King	40 kills

When fighter-bombers engage single-engine fighters deduct 20 per cent kills scored by fighter-bombers.

FIGHTERS ATTACKING SMALL BOMBERS

Note: When fighter-bombers are carrying bombs, they cannot perform the function as fighters. Losses are then according to this chart but bombs may be jettisoned (wasted) in order to act as fighter planes.

Per 50 fighters
Red: Kills as on card. Clubs: Double kills.
Spades: Treble kills.

Per 10 fighters
Red: Ace to 6, 1 kill; 7 to 9, 2 kills; 10 to Jack, 3 kills; Queen to King, 4 kills.
Clubs: Double above score. Spades: Nil.

Individual: Ace to 4, 1 kill.

Per 10 light bombers
Spades: as on card up to 6 (fighters shot down).
7, 1 kill; 8, 2 kills; 9, 3 kills; 10, 4 kills; Jack, 5 kills; Queen, 6 kills; King, 1 kill.

Individual: Black Ace, 1 kill; Red Ace, 2 kills.

FIGHTERS ATTACKING HEAVY BOMBERS

Per 100 fighters
Red: Kills as on card (max. 13). Clubs: Double kills (max. 26).
Spades: Treble kills (max. 39).

50 fighters: Count $\frac{1}{2}$ score (39 is 19).

Per 10 fighters
Spades and Clubs: 1 kill.
Hearts: Ace to 9, 2 kills; 10 to Jack, 3 kills; Queen to King, 4 kills.

Individual: Aces and King of Clubs, kill.

Per 10 bombers shooting
Spades: Kills as on card.

Individual bombers: Ace, 1 kill; King, 2 kills.

AIRCRAFT ATTACKING SHIPPING

Light bombers, torpedo-bombers, fighter-bombers with bomb load.

The chart is per 10 aircraft. Figures indicate number of hits. Figures in brackets indicate torpedo hits only.

Bombs: Hearts are complete misses.

Torpedoes: All red cards are misses.

Battleships and Carriers (C.V.'s)	*Cruisers*	*Destroyers*	*Smaller*	*Card cut*
1 (1)	1 (1)	1 —	—	Ace to 4
2 (1)	1 (1)	1 (1)	—	5 to 7
3 (2)	2 (1)	2 (1)	1 —	8 to 9
4 (2)	3 (2)	2 (1)	1 (1)	10 to Jack
5 (3)	3 (2)	2 (1)	1 (1)	Queen
6 (3)	4 (2)	2 (1)	2 (1)	King

Bomb score: 4,000 points (fighter-bomber 1,500).

Torpedoes score according to the Torpedo Effect Chart on page 138.

Individual aircraft attack (for odd numbers of aircraft)

Aces and 2's, bomb hit.

Aces of spades, diamonds and hearts, 1 torpedo hit.

BOMBING FACTORIES, PORTS, SHIPYARDS AND AIRFIELDS

Heavy bombers carry a 10,000-lb. bomb load. The load consists of ten 1,000-lb. bombs, each of which hits the target causes 1,000 points damage:

Ace to 3	1 hit per 10 bombers
4 to 5	1 hit per 5 bombers
6 to 7	1 hit per 2 bombers
8 to 9	1 hit per bomber
10 to Jack	2 hits per bomber
Queen	3 hits per bomber
King	4 hits per bomber

Light bombers (per 10 aircraft)
Hearts: 1 hit
Clubs: Ace to Jack, 2 hits
Queen to King, 3 hits

Each bomb is 1,000 lbs. Four are carried.

Aircraft attacking troops

I do not propose to include these rules because of adding unnecessary complication. If however, players conclude a 'World War' campaign and wish to start another, they may care to make rules for aircraft v. ground forces. Only allow smaller aircraft to attack army units.

Army group combat

The army groups are not supposed to recreate actual combat. That would make the game vastly complicated. Instead a system is used to merely inflict casualties when rival Army Groups engage in combat. The ground forces are used to simulate army operations without actually fighting according to real life. Army groups may engage in combat when 5 squares or less separate the contestants. Cut 1 card per 1,000 men remaining in army group. Casualties inflicted are 50 per card drawn, i.e. Ace equals 50 casualties inflicted. Card 6 equals 300 casualties inflicted and King equals 750 casualties.

When an army group (or whole army of several such groups) are outnumbered 3-1 or more a dice is cast.

1 to 3	surrender
4 to 6	fight on

This is done each move.

Army groups move at 5 squares per move or day.

A few Second World War ships made by Hansa and Delphin. Like all the other photographs of metal models in this book, they are scale 1–1250 or 1–1200 and made to a very high standard. They are rather too large for wargaming unless plenty of space is available. *Yamato* is 8 ins. long.

CHAPTER 9

The Hypothetical World War Game

In the author's opinion, land and air forces are necessary in order to wage true naval warfare. A sea battle that is staged merely for its own sake cannot be as rewarding as one that is fought because a vital issue is at stake. Sea power is an extension of land power. He who commands the sea can move his men and supplies.

The most rewarding game is that in which the player commands world power and deploys not only naval forces but land and air forces too. He plans grand strategy and master tactics and opportunities abound for evolving secret operations that may strike a major blow at the enemy or lead to disaster if they fail.

As a naval wargamer, your main interest will be in naval operations. Army and Airforce rules are kept as basic and simple as possible. Yet even so, they will perform the functions of armies and an air force and will be vital factors in moulding naval policy. The bombers can wreck the shipyards and destroy the vessels at sea. The army groups can occupy vital territory and force rival fleets to battle for the life links needed to support the troops. Sea battles occur only because there is a definite reason for the action taking place. If a commander wishes to place men on enemy held territory, he must first establish local naval supremacy.

This is a war of conquest. Your object is to rule the whole map and the most magnificent moment of the game will be when your army groups crash their way through the enemy homeland towards final and total victory.

Before that can happen, your enemy will have been crushed on land, sea, and in the air. It seldom happens at

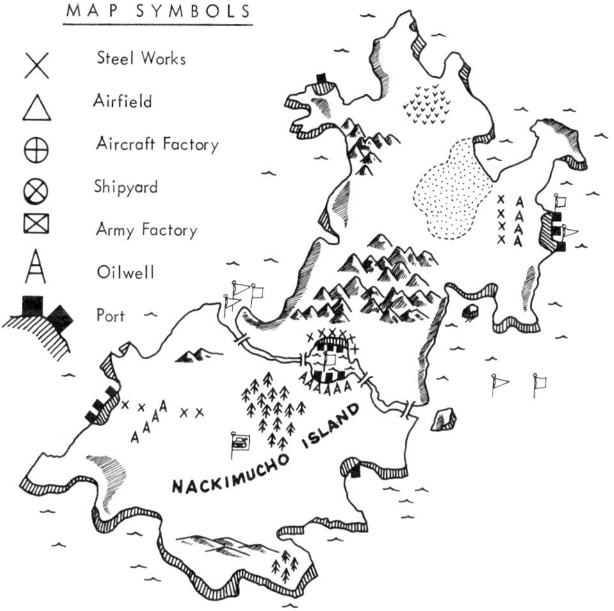

Drawing territories: An example of a territory on the 'World War' map. Nackimucho island. Located on the centre of the map. Some of the drawing is purely decorative, but woods, etc., can have a definite purpose in an advanced game (i.e. concealment factor for tanks and troops).

one blow. However, when one side at last establishes a definite superiority, it is usually the beginning of the end. Conquest means that raw materials formerly in his possession are now used to increase your forces. The enemy's war effort declines. The situation snowballs. Soon your power is overwhelming and your advance across the map quickens. The enemy is desperate. He can no longer offer serious opposition to massive invasion forces that occupy his territories in quick succession. Desperately he gathers his remaining forces together for one last major offensive. Unless he can pull off a master stroke, he knows his doom is just a matter of time.

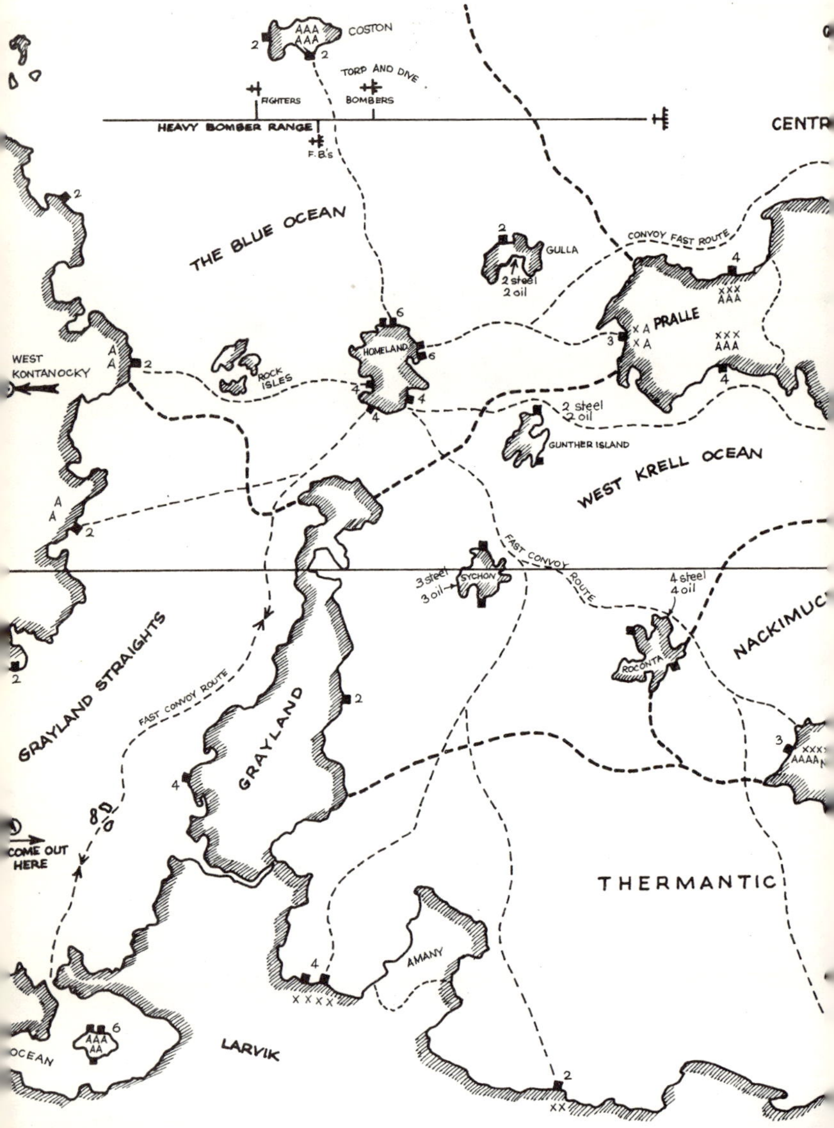
COSTON
TORP AND DIVE
FIGHTERS
BOMBERS
HEAVY BOMBER RANGE
F.B.'s
CENTR
THE BLUE OCEAN
GULLA
2 steel
2 oil
CONVOY FAST ROUTE
PRALLE
HOMELAND
ROCK ISLES
WEST KONTANOCKY
GUNTHER ISLAND
WEST KRELL OCEAN
FAST CONVOY ROUTE
SYCHON
3 steel
3 oil
4 steel
4 oil
ROCONTA
NACKIMUC
GRAYLAND STRAIGHTS
FAST CONVOY ROUTE
GRAYLAND
COME OUT HERE
THERMANTIC
AMANY
OCEAN
LARVIK

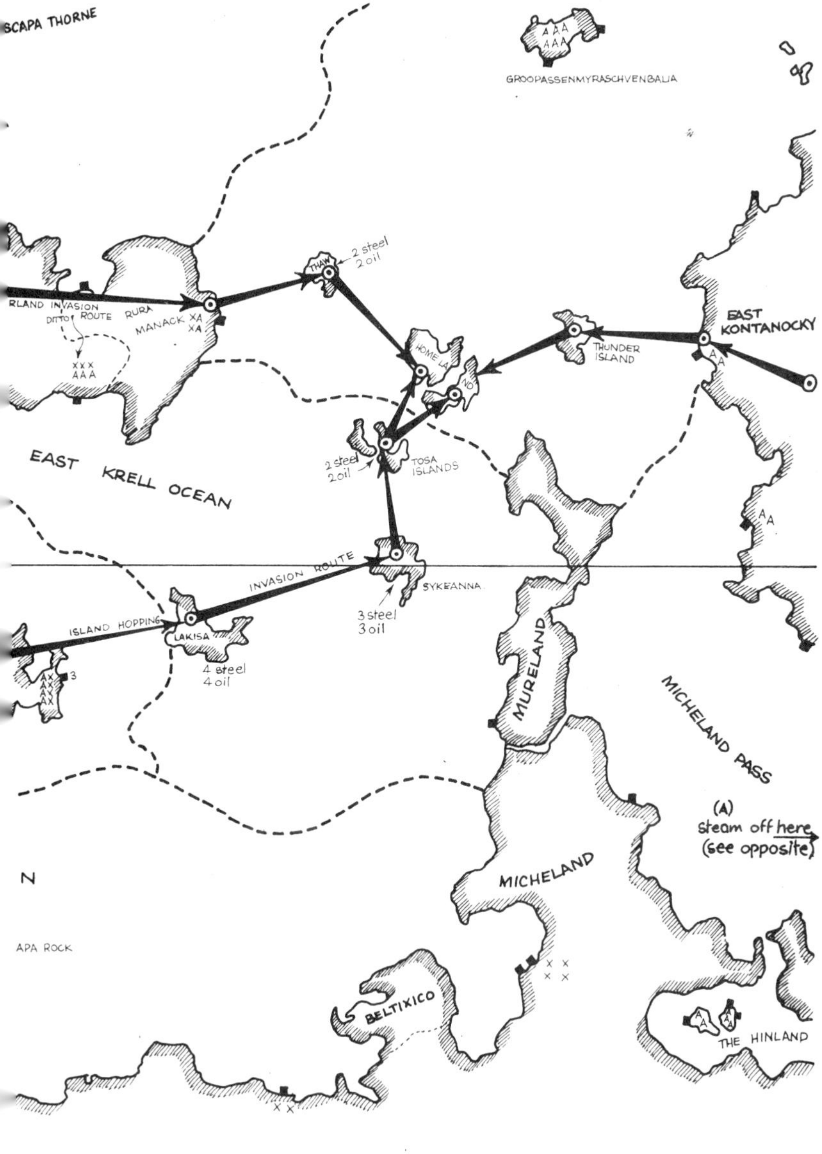
SCAPA THORNE
GROOPASSENMYRASCHVENBALIA
THAW
2 steel
2 oil
RLAND INVASION
DITTO
ROUTE
RURA
MANACK
EAST
KONTANOCKY
THUNDER
ISLAND
HOME LA
ND
EAST KRELL OCEAN
2 steel
2 oil
TOSA
ISLANDS
INVASION ROUTE
SYKEANNA
3 steel
3 oil
ISLAND HOPPING
LAKISA
4 steel
4 oil
MURELAND
MICHELAND PASS
(A)
steam off here
(see opposite)
N
APA ROCK
MICHELAND
BELTIXICO
THE HINLAND

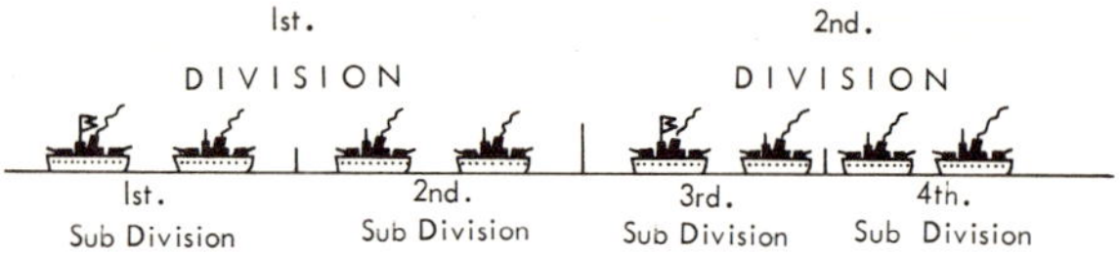

Structure of a Battle Squadron of Capital ships.

⚑= Flagship

The Squadron is commanded by a Vice or Rear Admiral
The 2nd. Battle Division is commanded by a Rear Admiral
The whole fleet of several Squadrons is commanded by a Vice or Full Admiral

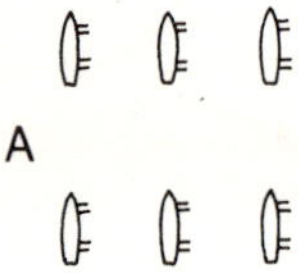

When not in action column or line abreast is the best formation. A torpedo is less likely to hit than if steaming in line ahead. The completely secret map move system makes this relevant to Battlegaming too.

If correctly spaced the formation (A) can quickly be formed in line ahead as shown. The outside column proceeds ahead.

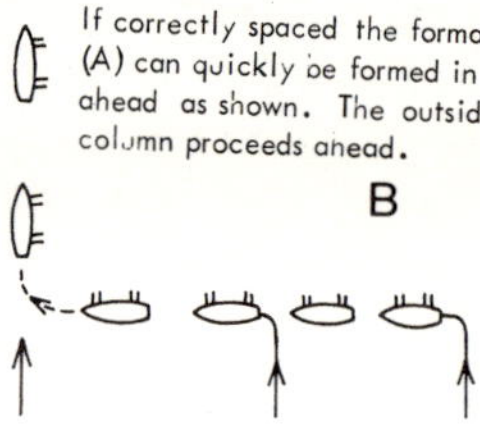

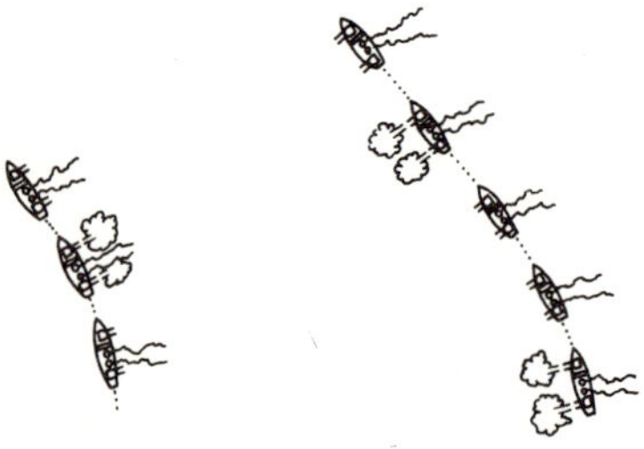

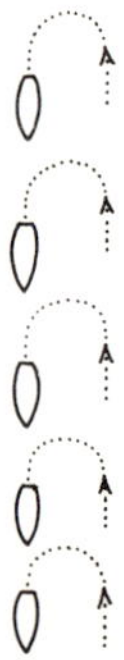

The larger fleet is faster and wishes to draw ahead of the three enemy battleships. The small fleet prevents this by circling and forces their opponent to expend its superior speed by steaming in a larger circle.

The famous German manoeuvre Gefechtskehrtwendung which involves turning a fleet through 180° in order to steam on an opposite course. It is a dangerous tactic because of the risk of collision. This is how the Germans escaped Admiral Jellicoe's terrible trap when he placed his fleet across their path as they advanced in line ahead. The normal tactic—to turn in succession—would have spelt disaster.

Note: This describes use of map on pages 146–147.

DESIGN FOR 4 SECTIONAL WORLD WAR GAME MAP

Smaller campaigns can be fought using a single section (i.e.) one sheet of 29 in × 21 in graph paper).

This map design is two-dimensional. A vessel, aircraft, or army group that moves off the left or right side of the map, will come out on the far side exactly opposite. See note (A) drawn on map.

Three of the most likely invasion routes to the enemy's homeland are shown from points exactly halfway between the two contestants.

By studying the map, it will be apparent that other routes are feasible.

If 4 persons wish to take part, 2 may jointly run one of the homelands or Amany and Beltixico may be used as two new homelands. The amount of ports, factories, and everything else applying to the original homelands will apply likewise to the fresh two. Homelands are named after commanders. If your name is Frederick you may call your homeland Fredland. These strange map names will soon 'grow' on you. After all, if you coined 'England' as a new name—it would sound very odd at first!

On small islands the number of oil and steel works has been indicated alongside and not actually drawn. These will of course be drawn upon the islands when the map is made.

Where ports are shown, they should be drawn on according to the number indicated against the port position shown on this diagram. Compare the enlarged drawing of Nackimucho with the diagram drawing. If a figure 5 is shown against a port position, it means that five black squares (docks) must be drawn to make up the port. Each dock is a blacked-in graph square.

Convoys fast routes have only been indicated for *one* of the homelands but compatible routes should be made for the other.

Obviously, if two people are going to fight such a game, then our World War must be an absolute mini-version. Neither can we reproduce its real elements in full. But we can reproduce the problems of supply and movement inherent in World War even if by doing so, we are obliged to make rulings that *simulate* the real thing. For example, rules contained herein for supplying steel are not very life-like but they do simulate the overall concepts of keeping our factories going in wartime. In the end it all relates to sea power. Unless the steel is safely transported home by sea, our building programmes will rapidly decline.

The campaign rules for this game can be either the 'Self Contained' game or the 'Semi-secret Game' both of which are described on pages 94–105.

It is suggested that the former is a more practical proposition until you have become well practised.

Perhaps the most unusual aspect of this game is that you do not command forces of any known nation. You rule your *own* country and your war is against another imaginary world power—that of your friend or brother!* The map design is hypothetical too and has splendid opportunities for trying out new and untried strategy.

It is far more satisfying to wage a global conflict against the personal armies, air-force, and navy of your pal, than it is to command (say) British or German forces. If you play this game, you are King of your own domain and can succeed where Hitler and Napoleon failed!

The basic workings of the game are simple. If a few friends join in, the systems can be made more realistic and elaborate, for then the increased complication can be handled by sharing out the paper work. This is definitely not the sort of game you can expect to finish in a day or two. It will keep you amused for weeks or even months. It can be modified of course so that a quick campaign is possible but if you are anything like the author, you will get as much pleasure from organising convoys and war production etc., as you will from actually fighting the battles.

Before the game can commence, it is necessary to draw

the battle map. The design of the map and instructions as to its construction are to be found in the preceding pages.

Selection of armed forces

Exactly equal forces are not desirable and would not be realistic. We therefore dice to determine the exact composition and strength of our armed services. If one side ends up by being rather more powerful than his opponent, then this is not really important. Most of the skilful opportunities available in this game are in the way operations and master strategy are conducted. A small variation in the initial size of the forces need not stop us beating the enemy.

Both players should throw one dice for every column listed below. In this way, the number of Battleships Heavy Cruisers, and so on that will comprise our fleet is determined.

The number of each type permitted by the dice throw must be written down for later reference and entry into the fleet list. When the strength of the fleet has been decided, we then cast dice for the number of various aircraft types that will form our initial air force.

Dice	*B'ship*	*H. cru.*	*L. cru.*	*Dest.*	*CV's*	*M.L.*	*D.E.'s*	*Corv.*	*Frig.*
1	6	6	9	15	4	2	5	5	5
2	6	6	8	14	4	2	5	5	5
3	5	5	7	12	3	1	4	4	4
4	5	5	7	11	2	1	4	4	4
5	4	4	6	10	2	1	3	3	3
6	4	4	6	10	2	1	3	3	3

Dice	*Sub.*	*Tank*	*Trans.*	*H. br.*	*Dive*	*Torp.*	*F.Br.*	*Fighter*	*A/Grps.*
1.2	4	12	15	10	5	5	10	12	15
3.4	3	10	13	8	4	4	8	10	18
5.6	2	8	12	6	3	3	6	8	20

Tankers and transport to a standard size of 10,000 tons.

Aircraft numbers above are squadrons. 10 heavy bombers to each squadron, 15 aircraft in other squadrons.

Each army group consists of 10,000 men.

Selection of warship types

The whole game is hypothetical but on the other hand we must use actual types of warships. Using playing cards it is

Ships leaving the homeland must not exceed 10 per day or move.

Example. *Ace—Bismarck* type *2—Yamato* class *3—Revenge* class etc. up to *King*.

Older types of ships should be included in each list. If a *Revenge* class battleship is drawn it could be employed on convoy escort or patrol duties.

The procedure should be repeated for each type. If you are allowed five battleships, select five battleship classes. It does not matter if three *Bismarck* class are drawn. They are no longer considered German ships so we can have a class of three or more!

The ships should be named and then entered into your fleet list. Remember that *you* are now King, President, Dictator or what have you of your own nation. The ships should be named by you because they are units of *your* fleet now.

Write down full details of each ship in your fleet list. These details are as follows. (1) Name. (2) Fleet number (3) Class. (4) Tonnage. (5) Armament details. (6) Speed. (7) Armour thickness. (8) Total points value.

Under each warship leave sufficient space to record damage and repair details. An example of a page from your fleet list is to be located on page 119. Do the same for aircraft and army groups. Each squadron and group should have a number and space must be provided to record losses and replacements.

At the start of the game a certain portion of each players fleet may be already at sea (Not transports). Throw I dice for each warship type.

Dice throw	*Large ships*	*Small ships*
1 to 3	None	None
4 or 5	1	2
6	2	4

Small ships: Destroyers and smaller.

The ships already at sea may be positioned anywhere up to 120 squares from a homeland port.

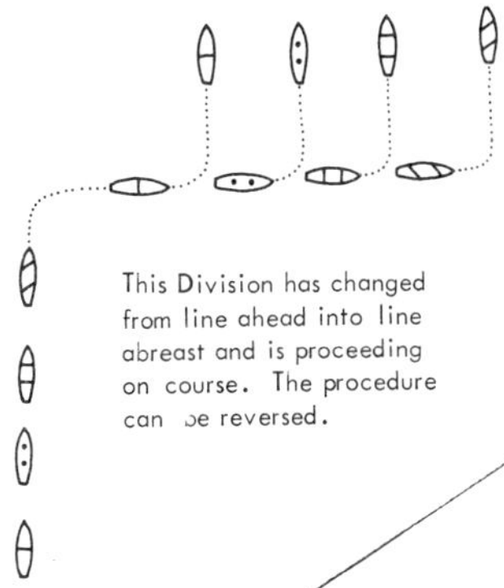
This Division has changed from line ahead into line abreast and is proceeding on course. The procedure can be reversed.

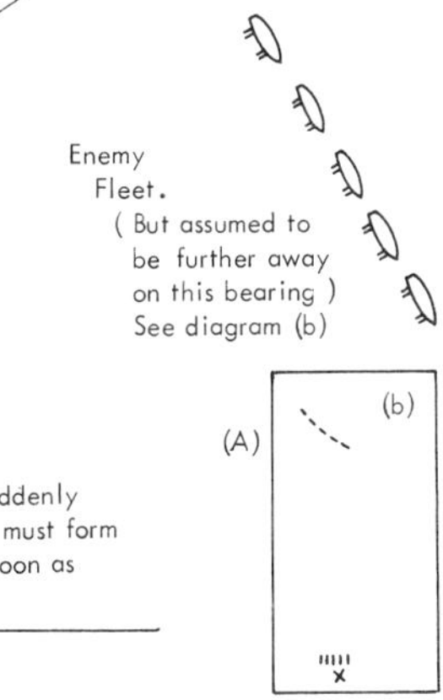
Enemy Fleet. (But assumed to be further away on this bearing) See diagram (b)
(b)
(A)
Enemy Fleet suddenly sighted. You must form line ahead as soon as possible.

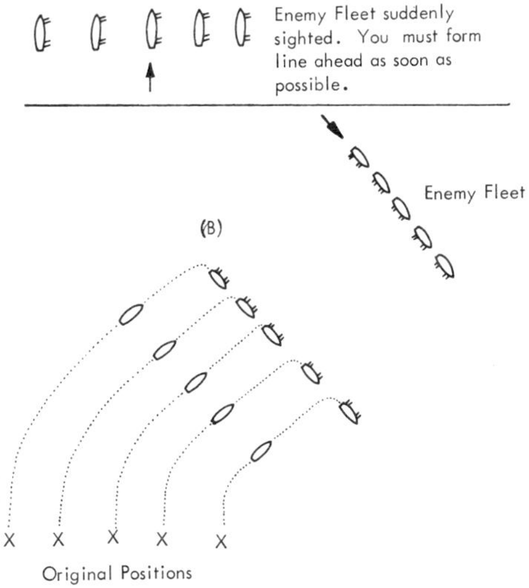
Enemy Fleet
(B)
Original Positions

Ships leaving the homeland must not exceed 10 per day or move.

Steel each steel works will supply 10,000 tons per week. When 10,000 tons has been taken from a steel works (which is the capacity of a transport) then the symbol should be circled lightly in pencil.

At the end of each week, all the circles are erased as the steel works can then supply a further 10,000 tons.

Steel is necessary for the following purposes.

(a) To build and repair shipping.

(b) To build aircraft.

(c) To repair factories or shipyards destroyed or damaged by bombing.

The points value of a steel works is 20,000.

One ton of steel will build one ton of shipping. Thus 10,000 tons of steel is needed to build a 10,000 ton ship but 500 tons per one in. of armoured belt must be added.

5,000 tons of steel is required to build an aircraft squadron (this figure is purely a convenient one!).

Airfield each side has five airfields upon his homeland and these may be positioned anywhere upon it.

An airfield can hold 2 squadrons of heavy bombers or 3 squadrons of other types.

Airfield value is 25,000.

Overseas airfields can only be built or repaired by Army Groups. An army group will take one week to build an airfield and can repair damage at a rate of 4,000 per day.

If aircraft remain upon an airfield when it is bombed or shelled then according to the percentage of destruction done to it will be a like percentage of aircraft destroyed.

Aircraft Factory Can produce 2 aircraft squadrons per week.

Points value 25,000.

Any factory or shipyard if damaged, will have production reduced in proportion to the damage done.

Factories can be repaired at a rate of 1,000 points value per day but 1 ton of steel per point lost is needed to effect repairs.

Countering torpedo attack

The turn-away. By steaming away from direction of attack, chances of hits being scored are remote. But contact with enemy battle-fleet is lost.

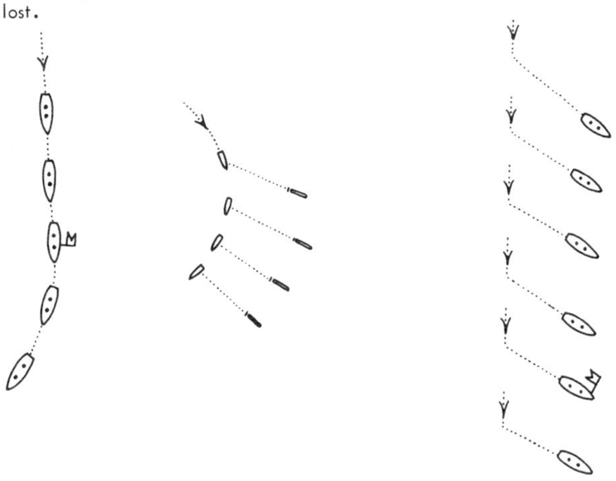

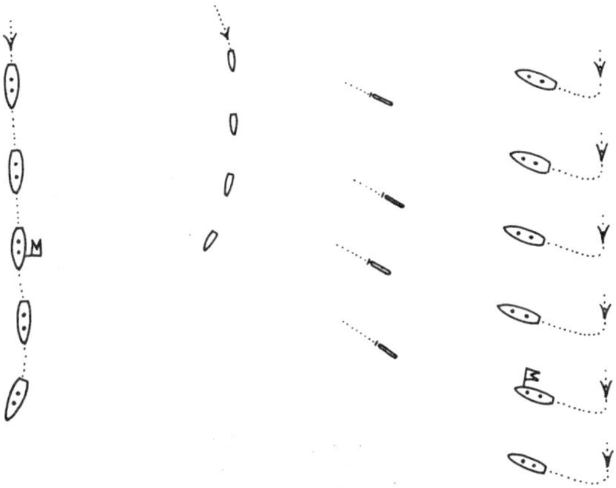

The turn-toward or 'Combing the torpedo tracks'

A much smaller target is presented but whole broadsides cannot now be fired at enemy battle-fleet which can make good practise in reply.

Shipyard Rules as for aircraft factory. Output as follows. 10,000 tons of shipping per week.

For the purposes of this game it is permissible to build a large ship utilising the building capacity of a number of shipyards together. The assumption would be that only one is actually building her!

New factories and shipyards can be built and the time taken is 25 days. Before work can commence 25,000 tons of steel must have reached the homeland and been allocated to the new factory or shipyard.

Each side starts the game with 5 factories and 5 shipyards and these must be positioned at least 10 miles apart from each other. There is also one army factory.

Army Factory This will turn out equipment and supplies for the army groups. 10,000 tons of steel a week are needed to keep it going. Points value and other rules are identical to Aircraft factories and shipyards.

Oilwell Each oilwell will supply 10,000 tons of oil per week. When the quota is taken from an oilwell, circle the symbol in pencil and erase at the end of each week. The points value of an oilwell is 10,000.

Oil is converted into units upon reaching the homeland. 10,000 tons of oil is equal to 50 units.

Port A port is the whole thing and each port consists of a number of docks. Each dock can contain one ship of any size.

A dock is represented by a single square that is blacked in. The value of each dock is 10,000 and when damage to a dock exceeds 5,000 it may not be used until repaired.

If a port is raided that consists of (say) 12 docks and 50,000 p.v. damage is inflicted by the raid, then five of the docks are destroyed.

Army groups can repair damaged or destroyed docks at a rate of 1,000 per day.

Stockpiles of army supplies are only destroyed if the whole port is wiped out.

Vessels may be bombed in port but it is not possible to torpedo them.

Transport and Tankers

As a point of convenience, all merchant vessels are of 10,000 tons.

Carrying capacity is 1 ton per ship ton. Thus a merchant vessel can carry 10,000 tons of steel, fuel, or supplies.

Vessels that sail along established convoy routes (marked upon the map) may steam at 5 knots above normal speed. This applies to merchant ships only.

To unload and load up at any port, merchant vessels must steam into port and not leave until the following move or day. This applies even if a vessel is but one square distant from a port.

Transport vessels can carry 5,000 men or one half of an army group. When carrying troops, a transport can unload upon the coastline as distinct from the normal procedure of having to call on a port.

A complete record of what is aboard each merchant vessel must be kept. When the ship loads and offloads a note of the shipment must be recorded in the relevant list or record.

Any cargo sunk at sea must be struck off the records. If aircraft or troops are aboard, the loss of same must be noted under the armed forces lists.

A transport can load I squadron of aircraft but not heavy bombers.

Warship Repair

Ships can be repaired at any shipyard or large port containing four or more docks. Repair rate as follows:

Battleships and Large Carriers	1,000 tons per day.
Cruisers and Small Carriers	500 tons per day.
Destroyers and Small Ships	100 tons per day.
Merchant Ships	400 tons per day.

Ship construction These repair figures also apply to the construction rate of warships. Before building can commence, sufficient steel to build the ship must have arrived at your homeland and been allocated to build her.

A 15,000 ton ship with an armoured belt of three ins. thickness, would need 15,000 tons of steel plus 500 tons per in. of armour in order to build her. The total steel required is thus 16,500 tons.

Out of every 10,000 tons of steel reaching the homeland, not more than 100 tons must be allocated to submarine construction and 500 tons to aircraft carrier construction.

Fresh ships built may be of any class desired and need not be selected by chance cards.

Loading and unloading at port Whenever a ship calls in port to offload or collect, it must put into that port and not off-load or collect until the following move. The vessel may then depart. When loading up, transport vessels need only put into the port nearest to where the goods are located. Troops must be moved directly to the ship.

Internment Any vessel putting into a neutral port for repairs-or to avoid destruction runs a slight risk of internment. The result of putting into that port is determined by a dice throw.

1. Ship not allowed to stay. Leave next day.
2. Two days' repairs allowed.
3. Four days' repairs allowed.
4. One week's stay and repairs allowed.
5. Indefinite stay permitted.
6. Ship siezed by Neutrals.

A neutral country is defined as territory which has not been occupied by either world powers.

Army Group replacement and supply

Two fresh army groups are available every week. These are on the homeland when first available and must be shipped overseas if it is desired to send them abroad. Army groups not receiving supplies may only fight at half strength and are not allowed to advance toward the enemy or to gain territory. If 3,000 men were left in a group that failed to receive supplies, then cards for casulaties inflicted upon the enemy may only be drawn for 1,500 men.

Supply rate is 5,000 tons of general supplies per group per fortnight. This must leave from a port in your homeland and may be deposited at a port nearest to the army group(s) concerned. It is assumed the group is supplied for a further fortnight when its quota has been unloaded.

Each operational overseas airfield must receive 10,000 tons of bombs and ammunition in order to remain operational.

Stockpiles General Supplies and bombs may be stockpiled at any port so that army groups and airfields can keep going as long as stocks last. 30,000 tons of bombs will keep an airfield going for three fortnightly periods for example. Stockpiles must remain at the point of unloading (unless re-shipped) and the fortnightly quota drawn off when needed. Not more than the proper quota can be issued until the fortnight is up. This is so that stockpiles can be overrun and captured or destroyed by bombing. (If the port is wiped out or the particular dock where the supplies etc. are stored).

Fuelling system for warships

The fuelling rules only apply to warships. We shall assume that merchant vessels are coal burners and that coal is readily available!

To run a realistic fuel system would not only be very complicated, it would also be very demanding on our time. This system is the least complex possible but simulates fuel problems quite well.

Cruising Speed: The cruising speed upon the map is 18 knots for all types of warships. If 18 knots is exceeded, fuel consumption is double the figures shown below.

Type	*Fuel units burned*
B. Ship.	5 per day.
H. Cruiser.	4 per day.
L. Cruiser.	3 per day.
Destroyer.	2 per day.
Smaller.	1 per day.

At the end of each move or day, simply count the numbers of each type *at sea*. Do not count those in port. Make a simple list as follows.

Date	*Battleship*	*Heavy cruiser*	*Light cruiser*	*Destroyer*	*Smaller*
1.10.39	3	4	4	12	18
2.10.39	3	6	3	12	19
3.10.39	3	5	3	15	15

Count large carriers as battleships and small carriers as heavy cruisers.

At the end of each weekly period, calculate the total fuel units burned and compare this with fuel stocks. Deduct accordingly. Those ships unable to remain at sea because there is not enough fuel to go around, must immediately put into the nearest port.

When oil reaches the homeland, it is converted into units (See *oilwell* page 156) 350 units are always available each week without the need to fetch oil from overseas. Oil from abroad is additional to the 350 units.

It should be observed that 350 units are only sufficient to keep a portion of the fleet at sea.

Fuel is not supplied direct to the ships because of the very complicated paper work that would be necessary. Once the oil reaches the homeland, it is simply allocated out to keep ships at sea or to enable others to put out.

The game is won when one side invades and occupies the enemy homeland. Occupation of the enemies capital city means he *must* surrender even if he still has forces available to fight.

If a player decides he cannot possibly avoid defeat he may sue for peace or surrender. The victor may reject peace terms and demand unconditional surrender. Failure to answer this demand means that the losing side is obliged to fight on.

Peace terms are pretty meaningless except in terms of humiliation. For example, the victor may demand terms as follows:

1. All warships over 10,000 tons to be handed over.
2. Compensation of £500,000,000 to be paid.
3. All overseas possessions to be ceded to the victor.

4. All generals, admirals, etc named by the victor as war criminals to be handed over for trial and possible execution.
5. Opponent to abdicate from power and no politicians may be put into office without victors approval.
etc.

You should keep a 'War Diary' or 'History of the War'. This is a day by day account of the war and really gives added spice and interest and can be read long afterwards to revive memories and nostalga on the game! Our own War Diary was filled with battle accounts, political speeches, propaganda, and charts and lists showing the progress of the war and comparisons of sinkings, output and so on.

Since the terms of any peace treaty would be entered into this war diary, it is obvious that an unkind victor may gloat about the harsh terms he inflicted upon his opponent whenever the contestants browse through the diary, or when perhaps, it is shown to interested friends!

This is a real incentive *not* to accept harsh terms and fight on, but if both sides wish to halt the game so that something else can be started then acceptable terms may be agreed upon.

CHAPTER 10

Battlegaming in Early Periods

This chapter is intended only to discuss naval warfare prior to the Napoleonic era for which rules etc, are provided in this book.

I shall not go into the matter in depth, for that is outside the scope of this book. I have however, provided the basic data for an 'Armada' battle but the reader himself will need formulate most of the rules-and research further. If research appears a formidable task, let me hasten to say that it is one of the pleasures of the hobby to dig for the facts and then apply them to your game.

For accurate details of the Great Armada enterprise. read the two books by Professor Michael Lewis. These are 'Armada Guns' and 'The Spanish Armada'. The latter is a paperback in the Pan series. The first named is a large book published by George Allen & Unwin Ltd.

I confess I have not seen the book 'Oared Fighting Ships' by R. C. Anderson but it is on my 'must' list and seems to be the bible of Galleys and Galleasses. The publishers are Percival Marshall. Another splendid book that naval enthusiasts will treasure is 'The Ship' by Bjorn Landstrom (Allen & Unwin) and the superb paintings and reconstructions of ships from earliest times to present are really invaluable to the model maker.

I make no apologies for the sketchy nature of this chapter or for the fact that I have not gone beyond the period of 1470 when the earliest guns first appeared on ships. My purpose is merely to point out the wargaming potential and hint at some of the exciting possibilities open to those who care to delve into the facts. To those whose appetites may be whetted and are eager to find out more. a visit to a big

A hastily assembled force of sixteenth-century ships, mostly made from small Airfix kits of *Revenge* and *Mayflower*.

library or to the National Maritime Museum at Greenwich will usually prove fruitful.

Ship fighting altered little between the 1650's and 1830 and so only somewhat minor modifications to your Napoleonic rules need be made to cover the period of the Dutch wars. It was a colourful era too. The warships were profusely decorated and surely the most attractive of all time to behold. The Dutch were a worthy opponent and proved one of England's toughest foes. Some really huge battles took place though these were never decisive in the sense that the Nile. Trafalgar. Tsu-Shima or Leyte Gulf were

.Once again the easiest form of research is to read books about the Dutch wars. It is not necessary to know all the technical details as long as the main details are understood. There are at least three large plastic kits of vessels around this period. They can be made into waterline models and single ship engagements can be staged. The three are *Sovereign of the Seas* and *Prince* (Airfix) Pyro plastics make the attractive French *St Louis*. I do not think scale is constant but this is hardly essential.

The German firm of Neptun make a few ships of scale 1-1250 in metal dating to the 1666 period. Otherwise. the battlegamer will need build his own models or else use ideas mentioned elsewhere in this book. Problems *are* involved. Problems exist so that men can overcome them!

Around this time the French fleet still consisted of a large number of Galleys. They appear to have been in use right up to the year 1717 or even later. A single sailing ship defeated nearly 40 galleys in the year 1684 but they did have certain advantages in calm weather. They have a colourful if not very useful role in a wargame, although it must be stressed that the battlegamer should call it upon himself to find out their shortcomings and the facts about them in some detail before writing his rules.

We next move backwards in time to the period of the war between England and Spain. I shall concern myself only with the Spanish Armada for which I think a highly interesting game can be staged.

Let us look at it from the battle side for now. The English were determined to settle the issue with gunfire. This was an astonishing and bold decision, for the gun was still though of as a weapon secondary to boarding. Philip of Spain realised the English intention and armed his Galleons with big cannons. The idea was to cripple the swift English ships in order that his troops might board. The English intended to use the long-barrelled Culverin so that they could destroy the Armada piecemeal from long range. We made a serious miscalculation here, because although the long guns could outrange the Spanish cannons, they could not cause any real damage except at close range.

Thus the gunfire of both sides was almost non-effective. A long gun can only, by nature of its purpose, fire a small ball. We found we could hit the Armada with impunity but that our shots were having almost no effect. And because our ships stood off when firing, Spanish gunfire failed to reach at all. Luck favours the bold they say and at least our tactics caused the jittery Dons to waste their ammunition in fruitless gunnery. Only when the Spaniards had run out of shot did our ships really dare to close the range and then the Spanish were in serious trouble. They had sailed down

Sailing ship tactics 'raking' the enemy from a position where he cannot reply.

This position is easier to attain when an enemy mast has been brought down.

the channel and reached their objective with hardly any loss. Having expended their ammunition doing so (wastefully) they now had to pay the price. The battlegamer who represents the Spanish side however, might not be so foolish.

What should be appreciated here is that the warships of this era could not sail anywhere near as efficiently as those of Nelson's time. The English Galleons were far handier than the bluff, cranky, beamy, leewardly, and unhandy Spanish galleons with their towering forecastles that made them such dreadful sailers. When in the grip of an unfavourable wind, the whole Armada was almost blown ashore and wrecked. Only a last moment change of wind saved them.

I have given here the details needed to resail a reduced Armada. The actual wargame rules must be drafted by the reader. It is possible to resail the *entire* Armada using the simple campaign rules contained in this book, under which rules, the use of models is dispensed with.

When landwargamers stage a battle, it is hardly practical to have 10,000 or so model soldiers taking part. They do like to stage large battles though and so they count 1 soldier as representing say 20 or 100. This enables the battle to be correctly proportional. The portion of grenadiers to guards for instance is correct.

I propose to do the same with the Armada. To reduce it to a scale that can be handled and keep the ship types present in proportion.

This is the result.

Spanish fleet	*English fleet*
5 royal galleons	6 royal galleons
2 great ships	24 auxiliaries
7 auxiliaries	
5 hulks or storeships	
6 light armed vessels	
25	30

Spanish galleons		*50-pdr.*	*32-pdr.*	*24-pdr.*	*17-pdr.*	*9-pdr.*	*5-pdr.*	*4-pdr.*
1. *San Juan,* 740 tons	(48)	1	2	6	8	5	5	6
2. *Rosario,* 740 tons	(48)	1	2	6	8	5	5	6
3. *S. Salvador,* 600 tons	(24)	3	3	5	6	1	–	4
4. *Annunciada,* 500 tons	(20)			10	–	8	2	–
5. *S.M. de Vison,* 400 tons	(18)			8	–	8	2	–
Great ships								
1. *La Regazona,* 1,200 tons	(36)	3	3	4	6	4	4	4
2. *La R. Coronada,* 800 tons	(24)	2	2	4	8	2	–	4
Auxiliaries								
3 ships of 400 tons ...	...		...	5	–	2	2	–
4 ships of 300 tons ...	...		...			2	2	2
Hulks: 200 to 600 tons each								
Light vessels (oars and sails) 50 to 100 tons (2 or 3 5-pdrs. and 4-pdrs. on each)								
Galleasses (one only)		4	4	8	–	3	7	–

ENGLISH SHIPS

Galleons		*32-pdr.*	*24-pdr.*	*17-pdr.*	*9-pdr.*	*5-pdr.*	*4-pdr.*
1. *Triumph,* 1,100 tons	(56)	4	3	10	14	16	2
2. *White Bear,* 1,000 tons	(52)	3	3	10	12	14	2
3. *Bonaventure,* 700 tons	(40)	4	4	8	12	6	2
4. *Revenge,* 500 tons	(30)			6	6	10	4
5. *Lion,* 500 tons	(30)			6	6	10	4
6. *Tiger,* 400 tons	(26)			6	6	8	2

Auxiliaries							
4 of 400 tons	–	2	4	10	6	–
4 of 300 tons	–	–	4	8	8	5
4 of 200 tons	–	–	–	7	10	4
4 of 150 tons	–	–	–	4	4	4
8 of 100 tons	–	–	–	3	3	3

When brackets are shown this indicates total number of guns. Surplus number above those listed are very small weapons—less than 4-pounders.

Guns carried are for the whole ship and *not* those mounted on each side.

SUGGESTIONS

Your rules should make Galleons a little more manoeuvreable than the auxiliaries. Spanish ships should not be able to sail very far off the wind direction. If 3 of the Spanish hulks are captured or destroyed, presume an English victory.

I shall include here one idea for a simple gunnery system. Allow each shot that hits to score its weight in points value against the ship's tonnage. This means that only 10 shots by the heaviest gun would be needed to sink a 500-ton vessel. We shall compensate for this by making shots difficult to hit.

At maximum range, turn 1 card per shot fired. Ace is a hit.

At half range, Ace or 2 is a hit.

At point blank range (one-fifth of maximum range) Ace to 4 is a hit.

These rules, simple though they are and open to improvement though they be, will serve to encourage English vessels to keep their distance. This will be true of actual tactics.

The Galleasse should not perform well under sail but should have a fair degree of manoeuvreability under oar.

Gun	*Range*	
50 pdr.	1,000 yds.	
32 pdr.	850 yds.	
24 pdr.	800 yds.	
17 pdr.	1,250 yds.	Long Guns or Culverins.
9 pdr.	1,250 yds.	
5 pdr.	850 yds.	
4 pdr.	800 yds.	

A certain number of moves can be allowed in order for the 'Armada' to try and reach the Duke of Parma's army and then sail to invade England.

I have of course, merely touched the surface of this 'game' and the reader who wishes to do this will need to spend a few evenings working out the details and rules. It is all entirely practical and should prove a very entertaining game. The ships do not have to be models. They can be 'flats' on stands or else ship shapes (wood or cardboard) as seen from above. Although the English will have more ships, their task will be a difficult one, for to close the range will make them vulnerable to Spanish heavy cannon fire. Your rules should allow for disabling hits and the fact that the Spanish Galleons and Auxiliaries (the two Great ships were big auxiliaries) had well over 18,000 soldiers aboard. These troops were reckoned to be first class fighting men. The English fleet had less than 2,000 soldiers. The English could probably fire faster than their enemy, for the Spanish had no real love for the gun and probably did very little training.

The rules should also make English fighting vessels fast, manoeuvreable (compared to the Spanish), and able to sail further off the wind direction.

Seventeen years before the Armada sailed, there occurred a tremendous battle between a Christian fleet and the fleets of Turkey and Algeria.

The bulk of the Christian fleet was Spanish but there were also contingents from the republic of Venice, Malta, Genoa, and a force provided by Pope Pius V. The battle is known as the battle of Lepanto and nearly 500 galleys were involved. The Venetian force included six Galleasses which because of their tremendous fire power, played a major part in the fight. They mounted 10 heavy cannon and 12 or so smaller ones, whereas the ordinary galley usually mounted two guns firing forward.

The two fleets advanced and fought like armies. The Christian fleet was better equipped but numerically inferior. I am going to suggest that this is a really practical and colourful battle to restage. What is interesting is that whereas the actual battle was little more than two armies meeting at sea, all kinds of fascinating things have to be taken into account. Gunfire, ramming, boarding, and small arms. Christian small arms consisted of the somewhat primitive arquebus and sword and shield. A certain degree of armour was also worn. The Turks used the scimitar and bow and arrow. It was truly a tremendous battle with about 150,000 men taking part (a large portion were oarsmen). The Christian losses were 24,000 killed and wounded. The Turks are thought to have lost about 20,000 men and a further 50,000 wounded.

To fight a mini-version of this battle, or one like it, would not be terribly difficult. About twenty models would need be made but these do not have to be masterpieces in order to have a good game. Once again a short cut is possible by merely drawing the vessels, mounting them on card and stands. Once boarding takes place, it may be possible to adapt some well thought out land wargaming rules. The first step to take will be to read about the battle of Lepanto.

The next period of great interest is the early part of the 16th century. Here we are introduced to vessels like *The Great Galley* and *Henry Grace a Dieu*. The former, as its name implies was a huge Galley. She had nine sails, 160 oarsmen, and the staggering total of 210 guns!

We should not be deluded by her impressive sounding armament, for guns of this period (1515) were very primitive and of the bombard type. In any event, most of her guns must have been railing pieces.

The *Henry Grace a Dieu* or *Great Harry* is better known than this great galley. She carried 186 guns but broadsides from ships of this era were not likely to cause much material damage. Most of them fired stone shot and the guns must not be thought of as cannons as we know them.

The vessels though are extremely impressive. Such ships as *Great Harry* were floating castles that towered out of the water. They had huge forecastles that made them crank and leewardly. They must have been very bad sailers and it is said the *Great Harry* never ventured into the open sea. Pyro plastics make a nice 10 in. long model of the Portuguese 140 gun *Santa Catarina do Monte Sinai* which is very similar to the English ship. The model is festooned with flags and shields and gives a good visual effect of the clumsy design. A prime feature of this design was to ward off boarders and that is why they were so high and awkward. I have made my model of the Portuguese ship into a waterline model. By putting together two of these ships and altering one of them to represent something else, an impressive single ship action can be fought. The big ones weighed nearly 1,000 tons and had a crew of 700 men (250 soldiers included). The fighting men wear armour and use bows, pikes, swords, and possibly spears.

Scottish readers may be interested to hear that the king of Scotland had built in France the *Great Michael* which was another of these monsters. It would be somewhat amusing to make models of her and her English rival and then sail down (or up!) to fight it out. To make it more interesting the issue at stake could be something that as a matter of pride would make both contestants really determined to win.

Models of both *Great Harry* and *Santa Catarina do Monte Sinai* could be made from drawings and paintings in the book 'The Ship'.

The earliest period that we can fight a gun action is around 1470. The carrack's of this age were similar to the well known *Santa Maria* in general design although perhaps a little more primitive.

A splendid reconstruction of one of these ships is to be found in Bjorn Landstrom's book 'The Ship' and shows a vessel armed with five small guns on each side of her aftercastle. To fight an action using models of ships such as these would at least be an historic occasion, for it marks the beginning of the age of the shipborne gun. Exactly 500 years later, it is interesting to see that the gun is still with us—albeit that its role is secondary to that upstart the missile!

Earlier periods than this offer a fascinating range of ships and tactics that are open to the wargamer. If large ships are of interest then we cannot fail to be intrigued by the huge fighting vessels of around 200 B.C. one of which is known to have needed 4,000 oarsmen.

Some of the ancient vessels were fitted with wierd war engines and it seems that Greek fire and quicklime and other such novelties can keep the battlegamer amused for very long periods. Rewarding though it must be to delve into some of the facts and figures, I have desired to limit this survey to the age of gunfire. Perhaps the reader will care to investigate the battlegaming potential where I have left off. It really does not matter whether you use 18-in. guns or rock throwing war engines, battleships or galleys. Whatever it is that manoeuvres on the sea, wargamers will be well rewarded when they turn back the pages of history and repeat the efforts of long dead Admirals.

CHAPTER 11

A Method of Skill

The first wargame I ever took part in was a simple affair that a few friends and I staged while serving in the R.A.F. back in 1954.

A flight of imaginary bombers (no models) had to reach and bomb a target. So many moves were allowed and the other side had a specified number of fighters with which to attack the bombers. Those that survived fighter attack then bombed the target and the winning side was determined by the damage done.

It was a simple affair but it was dependent on a certain amount of skill as we threw darts for combat and bombing results. The enjoyable part was when, as a result of our skill at dart throwing, we were able to shoot an enemy bomber out of the sky or else tot up an impressive bomb damage score.

Now it may be said that throwing darts do not have much in common with the real thing (i.e. firing machine guns). Of course this is true but it is true of whatever method is used short of both parties starting their own airforce! Casting dice and turning cards is of course even further detached from realism, because darts at least have to be aimed and sent through the air.

It is my opinion that darts represent the best method I know of introducing an element of skill in firing guns. I confess that my own preference is to rely on tactical skill to win the game and trust to luck when shots are fired by using cards and dice. After all, everyone is trained in war but one wargamer may be vastly more practised in the method of skill (whatever it is) that is being used. Also, battlegaming is a hobby intended to put everyone in the picture. Someone who just does not have the flair to perform

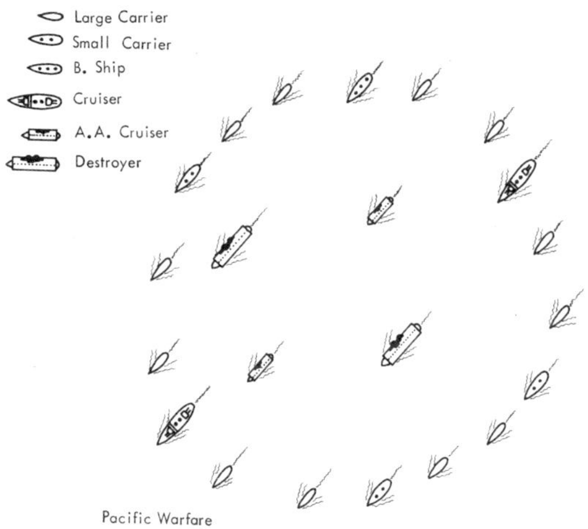

Pacific Warfare
American Task Force   Air Defence Formation

a certain skilful operation is going to be left out.

Still, it *is* true that in real life skill counts for an awful lot. Lack of training often leads to a decline in performance too, so there may be something said for reproducing this element provided one player is not so vastly better than another that he cannot fail to win every time. That is not true to life whatever nations are represented—at least as far as the age of steam goes.

The idea of guessing the range from one model to another is well known. The only trouble is that the method ignores one important factor. That is the increased accuracy of larger guns.

Ken Fisher got over this problem by absolving the need to estimate the fall of shot of each gun. Instead, one range was called out to represent the whole salvo. If there was an error of 3 ins. this would still be a hit or hits as far as a 12-in. gun goes. A 6-in. salvo would need be within $1\frac{1}{2}$ ins. to score possible hits and so on. Once a salvo was correctly placed a card was drawn to see how many hits were obtained.

A common objection to the use of darts might be that one player is no good at it or else has never thrown them before. If there is a too obvious advantage for one player. then let him command a smaller force to compensate for this. After all, rival fleets are seldom equal.

The interesting thing about using darts is this. Our skilled player might have a sudden run of bad luck or bad form. It happens to us all. The poor player might suddenly pick up and surprise his opponent with a burst of good throwing. This happens too. And even a wildly thrown group of darts, can often by pure chance, score very high. In fact the apple-cart can be upset in a number of ways.

Now observe another very interesting fact. Poor morale can very often cause a marked decline in rate of fire and accuracy. Observe Tsu-Shima and aspects of Jutland for example. Now this becomes self evident when throwing darts. If it is absolutely vital we make a high score the dread of doing badly induces the darts to go other places than intended. The more anxious, depressed, or over-excited we are, the more likelyhood there is that our shooting will reflect this. Cool calm confidence will tend to help our scoring but anything is likely to happen to put a buoyant thrower into a sudden mood of anxiety or depression.

If you do decide to use a dart throwing method of gunnery or air fighting etc. I do suggest that morale rules will not be necessary!

Now let us see how this idea can be applied to naval battle-gaming.

We simply make the dart take the place of the dice. Observe the damage inflicted chart contained in the simplified set of Napoleonic rules. It indicates a damage percentage of a certain figure from one quarter of this figure to times four. Here is how we do it with darts

Score (3 Darts)	*Damage percentage*
under 30.	$\frac{1}{4}$ only.
30 to 40.	$\frac{1}{2}$ only.
41 to 50.	$\frac{3}{4}$ only.
51 to 60.	Normal.

61 to 80.	× 2
81 to 100.	× 3
Over 100	× 4

This is one example only but it should be obvious that similar charts can be compiled to cover any situation.

Dartboards are not expensive things to buy if we select a model made of cheaper materials Even toy shops sell inexpensive dart boards that are quite good enough for our purpose even though they are rather small.

I think even the non darts player will find it an enjoyable affair. He will quickly pick up the art and may develop into a better shot than he thinks possible. Even if the *game* of darts does not appeal to him in the slightest. he may find that doing it for the sake of blowing his opponents forces into eternity is rather more rewarding!

The classic naval manoeuvre. Placing one's fleet across the path of an approaching enemy. The head of each enemy line is being shattered and he is unable to return any fire of note. Line (B) is trying desperately to form line ahead parallel to the enemy but each ship, as it turns, is subject to concentrated fire.

Line (A) and (B) may be said to represent Nelson's attack formation at the Battle of Trafalgar. This enabled him to break the French and Spanish line and shatter it. Such tactics could not have worked against a fleet well practised in gunnery.

At Tsu-Shima in 1905, Tojo turned his fleet in succession within range of the Russian fleet (as in line B). This was to put his fleet on a favourable heading and he calculated that Russian gunfire would not be accurate, thus making the risks involved acceptable.

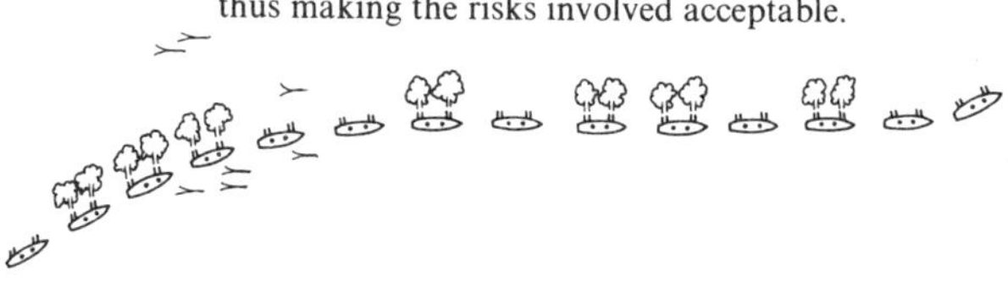

REFERENCE BOOKS

Reprints

Janes Fighting Ships for the years 1898, 1906, 1914 and 1919.

Weyers Taschenbuch der Kriegsflotten 1905, 1914, 1918 (all nations). Contain simple plans, complete technical data, gunnery tables etc.

Modern books

Die Deutschen Kriegsschiffe 1815-1945. Volumes 1 and 2. Contains elaborate plans and complete technical data of all Germany's warships.

Complete set of books (battleships, cruisers etc.) published by the historical department of the Italian Navy. Plans, sectional drawings, technical data. Superb works of reference. Covers mostly Italian ships. Books can be purchased separately.

Battleship book called 'Le Navi di Linea Italiane 1861/1961'.

British Battleships by Dr Oscar Parkes. Superb work of reference that took over a quarter of a century to compile. Plans and technical data. May not be currently available. Also 'British Destroyers' (similar).

Deutsche Kriegsshiffe in zwei Jahrhunderten three volumes. Small booklets covering Germany's warships during the age of steam. Technical data and simple plans.

Semi-Technical Books

Limited number of plans (if any) and general ship data. List all ships in fleet concerned.

Osterreichs Kriegsmarine 1848-1919. Austrian Fleet.

The Wooden Fighting ship in History. British wooden ships from earliest times to late 19th century.

These are just a small selection of available books. A visit to the bookshops that sell new and secondhand naval books are well worth a visit.